Biophysics and biochemistry at low temperatures

Freeze, freeze, thou bitter sky,
That dost not bite so nigh
    As benefits forgot:
Though thou the waters warp,
Thy sting is not so sharp
    As friend remember'd not.

As You Like It
William Shakespeare

Biophysics and biochemistry at low temperatures

FELIX FRANKS

Director, Cryopreservation Division, Pafra Ltd, Cambridge
and Senior Research Fellow, Department of Botany, University of Cambridge

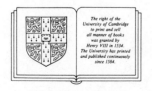

The right of the
University of Cambridge
to print and sell
all manner of books
was granted by
Henry VIII in 1534.
The University has printed
and published continuously
since 1584.

CAMBRIDGE UNIVERSITY PRESS

Cambridge

London New York New Rochelle

Melbourne Sydney

Published by the Press Syndicate of the University of Cambridge
The Pitt Building, Trumpington Street, Cambridge CB2 1RP
32 East 57th Street, New York, NY 10022, USA
10 Stamford Road, Oakleigh, Melbourne 3166, Australia

First published 1985

Printed in Great Britain by the University Press, Cambridge

Library of Congress catalogue card number: 84–21474

British Library Cataloguing in Publication Data
Franks, Felix
Biophysics and biochemistry at low temperatures.
1. Cryobiology   2. Cold adaption   3. Cold –
Physiological effect
I. Title
574.19′167   QH324.9.C7

ISBN 0 521 26320 4

Contents

Acknowledgements

It is a pleasure to acknowledge my debt of gratitude for various services rendered, including lessons in simple biochemistry and plant physiology, reading various parts of the draft manuscript, making available results prior to publication and discussing various contentious issues: Tom ap Rees, Ian Woodward, Patrick Echlin, Peter Lillford, John Morris, Harry Levine, Louise Slade and John Blanshard.

It is a special pleasure to thank my wife Hedy who, as on so many previous occasions, provided sterling help with important tasks such as checking and typing the bibliography and proof reading.

Foreword

Cold is the fiercest enemy of many forms of life. This is due partly to the general slowing down of physiological processes at suboptimal temperatures, but mainly to the fact that the essential chemical of life – water – happens to freeze at a temperature which is widespread in the ecosphere. The freezing of tissue water and the resulting freeze concentration of all soluble matter can have devastating consequences, unless the organism is properly prepared to resist the physiological stresses.

There exists a vast literature describing the symptoms of cold injury and the metabolic and morphological changes which various forms of life undergo during the cold hardening period. The connection between injury and survival on the one hand, and the temperature induced changes in the aqueous substrate on the other, is seldom considered. The physical properties of water are extremely sensitive to changes in temperature and changes in the concentrations of dissolved solute species. Such sensitivity may well be amplified in the responses of biological structures and life processes to changes in the hydrogen bonding patterns that exist in water.

During the past six years my colleagues and I have been engaged in studying the responses of *in vitro* and *in vivo* systems to low temperatures, and we have come to realize the interplay of a wide range of principles and disciplines: the mysteries of undercooled water, the *in vivo* nucleation and propagation of ice, both spontaneous and catalysed, cold labile proteins, the properties of concentrated aqueous solutions, especially those of carbohydrate origin, cryobiochemistry, biological antifreezes and biogenic nucleation catalysts, cold hardening mechanisms, laboratory cryobiology, cell membrane energetics and dynamics, and others.

I thank the cloud physicists who taught me about undercooled water, the metallurgists who taught me about nucleation in condensed systems, the haematologists who explained the intricacies of the red cell membrane, the insect physiologists who cleared up my misconceptions about the developmental stages of insects and the many others who helped me to synthesize my own ideas about causes and effects in low temperature injury and resistance. I thank them all.

The approach adopted in this book is that of one trained as a physical scientist who drifted into the life sciences fairly late in life and never received any formal teaching in biological dogma. The vocabulary is hard to assimilate and even harder to memorize. The book developed from a lecture course which I gave in the Department of Botany several years ago. It is by no means a comprehensive account of the subject. Its purpose is to sketch out the overlapping areas and disciplines where the interested student must search for solutions to the many unresolved problems.

Cambridge, 1985

1

Water, temperature and life

1.1 Low temperature in the ecosphere

Before entering into a discussion of the effects of low temperatures on life processes, it must be emphasized that the concept of *low* temperature is a relative one. Thus, to a metallurgist, a low temperature might be one at which liquid metals are quenched and tempered, whereas to many a physicist a low temperature is one associated with quantum effects such as superfluidity and superconductivity. The two temperature ranges differ by some 800 degrees. Within the context of this book, low temperature is defined arbitrarily as the suboptimal temperature range over which living organisms can function or at least survive in a state of dormancy, or the range of temperature employed in the laboratory to preserve biological material in a state of suspended animation. In broad terms this temperature range can be subdivided into three regions: at $> -20\,°C$ metabolism can still take place and a cold adapted organism is usually in an undercooled (unfrozen) state; -20 to $-80\,°C$ is referred to as the anabiotic state which is critical for life. The water in the organism is frozen, unless special protective mechanisms are at work. Temperatures below $-80\,°C$ are very rarely encountered in the natural terrestrial environment, although they are commonly employed in the laboratory preservation of biological material. Thus, the lowest temperature considered in this book is that of melting nitrogen, $-210\,°C$ (63 K), a substance which finds application as a cooling medium for the fast quenching of biological specimens for electron microscopic investigation. Apart from those chapters that deal specifically with the laboratory application of low temperatures, we shall confine ourselves to physiologically suboptimal temperatures that occur in the natural environment.

1

1.2   Water and life

The intimate involvement of water in life processes is generally taken for granted to the extent that the role of water in biochemical and physiological events is ignored or, at best, oversimplified (Franks, 1982a). There are very few metabolic or biosynthetic reactions in which water does not play a part either as reactant or product. The four major types of biological reactions involve oxidation, reduction, condensation and hydrolysis, with water playing a major role in each. Thus, photosynthesis and transpiration of plants involve an annual water turnover of 10^{11} tonnes, and the net daily water turnover of the normal human adult amounts to no less than 2.5 kg, equivalent to 4% of the average body weight (Franks, 1981a). Water is also intimately involved in the maintenance of macromolecular and supermolecular structures, such as those formed by combinations of peptides (Franks & Eagland, 1975), nucleotides (Eagland, 1975), carbohydrates (Suggett, 1975) and lipids (Hauser, 1975). At the macroscopic level water is the transport medium which carries a supply of nutrients through the organism and which removes waste products. Finally, water is the life-long environment of many organisms. Those, like mammals, which have evolved to survive on 'dry' land (really dry land does not exist), have done so at a considerable cost through the development of mechanisms to control their water balance and avoid desiccation. However, even such higher forms of life still begin their development in an aqueous medium.

Where account is taken of the role of water in life processes, this is commonly done by the assignment of a limiting water activity (a_w) above or below which the particular organism cannot function (Gould & Measures, 1977; Leistner, Rödel & Krispien, 1981). For most organisms the optimum condition for proper functioning corresponds to $a_w = 1$, so that $a_w < 1$ constitutes a physiological water stress condition which increases in severity as a_w decreases. The relevance of a_w to physiological function and well-being is open to doubt (Franks, 1982b) and will be analysed in more detail later in this book. At this stage it is sufficient to emphasize that a_w is a thermodynamic quantity which, in a system at equilibrium, is by definition proportional to the partial vapour pressure of water, a quantity which is itself proportional to the concentration of solutes and does not depend on their chemical nature. Furthermore, the standard state of unit activity is pure water at any temperature, so that $a_w = p$(water in solution)/p(pure water), independent of temperature. Since, however, the vapour pressure of water is itself a function of temperature, any comparison of a_w values at different temperatures must take this into account.[†]

† The various assumptions and limitations which form the basis of the activity concept (Lewis & Randall, 1961) need to be clearly understood and appreciated

The physical properties of liquid water are unique in many respects and they are also very sensitive to changes in temperature. It is likely, therefore, that the marked effects of temperature on biochemical and physiological processes are related to these changes in the physical properties of liquid water. The physical basis for the eccentric behaviour of liquid water is the structure of the H_2O molecule itself (Kern & Karplus, 1972). To a good approximation the molecule can be regarded as a sphere of radius 0.282 nm which has embedded in it two positive and two negative charges, placed at the vertices of a regular tetrahedron, as shown in Fig. 1.1. The positive charges can be identified with the positions of the two hydrogen atoms and the negative charges with those of the two lone pairs of electrons in the electronic structure. The H_2O molecule thus has the ability to form four hydrogen bonds, with two proton donor and two proton acceptor sites. The spatial disposition of these sites gives rise to a three-dimensional hydrogen bonded network, with each oxygen atom surrounded by four other oxygen atoms and with a hydrogen atom placed on each O—O axis. Ordinary hexagonal ice is the ideal manifestation of such an arrangement. In the liquid state thermal energy opposes the structural forces responsible for this ordered arrangement, and the question is to what extent such ordered molecular arrangements can persist in the liquid, bearing in mind

Fig. 1.1. The four-point-charge model of the water molecule. The oxygen atom is placed at the centre of a regular tetrahedron, the vertices of which are occupied by two positive (hydrogen atoms) and two negative (lone electron pairs) charges, the O—H distance being 0.1 nm. The distance of closest approach of two molecules (van der Waals radius) is 0.282 nm.

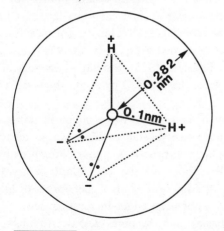

before a_w is cited as a criterion of physiological function or technological quality. Frequently the relationships are not nearly as obvious or clear cut as the physiologists or microbiologists would have us believe.

that the hydrogen bond must be considered as a weak interaction compared to a covalent bond. As will be seen in Chapter 2, extensive hydrogen bonding appears to be maintained in the liquid state, at least on a time averaged basis. On a short-term, picosecond time scale, diffusion takes place, but the preferred atomic sites remain those in which a distorted, but essentially tetrahedral, network is maintained (Rahman & Stillinger, 1973). This arrangement of water molecules in the liquid state is normally referred to as 'water structure'. Being dependent on weak forces, such a structure is easily perturbed by temperature and solutes of various kinds. In turn, water is able to modify the interactions between solute molecules and within solute molecules in the case of macromolecules. The role of water in controlling biological processes stems from such weak hydration interactions and thus from the intimate details of the water–water hydrogen bond.

Striking evidence for this conclusion is provided by the effects of D_2O substitution on life processes. In terms of structural changes in the liquid, the substitution of H by D is expected to produce only minor perturbations. This is borne out by comparative diffraction studies on the two substances in the solid state (Hobbs, 1974). The intramolecular bond vibration energies are of course affected, due to the difference in zero point energies of the H_2O and D_2O molecules, but the intermolecular (hydrogen bond) vibrations are not subject to zero point energy. There is little reliable information about the comparative strengths of the hydrogen and deuterium bonds, although it is usually assumed that the O—D—O bond is somewhat stronger than the O—H—O bond (Frank, 1972).

Thus, D_2O has the higher melting point and boiling point, although in the ice lattice the two bonds appear to have the same length. Chemical reactions performed in the two solvents exhibit kinetic isotope effects, such that reaction rates in D_2O are retarded by approximately 30% compared to those in H_2O (Frost & Pearson, 1953). Similarly, the self-diffusion coefficient of D_2O is lower than that of H_2O by about the same amount (Kell, 1972).

All these differences may be considered to be minor in terms of the marked similarities between the two substances. However, as regards the solvent involvement in biochemical processes, major differences become apparent. Only the simplest forms of life can adapt to total D_2O substitution, and then only gradually. For all higher forms of life D_2O produces the symptoms of a water stress which increases in severity as the level of substitution increases and becomes lethal at substitutions approaching 40% (Franks, 1982a). The effects of D_2O substitution can be demonstrated on isolated biochemical functions, such as cellular ATP production, membrane enzyme activity, protein synthesis or electric excitability. In every case it is found that D_2O has an inhibiting effect which increases with increasing degree of substitution. Superficially the effects produced by D_2O resemble those accompanied by a reduction in temperature, but detailed study reveals rather more subtle effects (Frank, 1972). While their discussion is beyond the scope of this book, the D_2O substitution results suggest that a low a_w can hardly be the primary cause of physiological water stress symptoms, notwithstanding the fact that a_w often provides a superficially good correlation with physiological viability.

1.3   The perturbation of water by solutes – hydration interactions

The structure of liquid water, based as it is on the approximately tetrahedral distribution of molecules, is easily perturbed by solutes. The nature of such perturbation depends on the type of solute particle; three major types of hydration interactions can be distinguished, according to the manner in which the intermolecular hydrogen bonding in water is affected. An extensive literature exists on the energetics of ion hydration (Friedman & Krishnan, 1973; Conway, 1981), and recent advances in neutron diffraction techniques have added fresh impetus to structural investigations of the ionic hydration shell in solution (Enderby & Nielson, 1979). Bearing in mind the molecular structure of the water molecule (Fig. 1.1), the geometry of the hydration shell is dictated by ion–dipole forces, as shown in Fig. 1.2. It is of interest that the essential features of these hydration geometries appear to be identical for Ni^{2+}, Na^+, K^+, and Ca^{2+},

Fig. 1.2. The hydration shells of monatomic cations (*a*) and anions (*b*), as determined from neutron scattering. The ion–oxygen distance and angle of inclination θ and ψ are obtained indirectly from the intensities of the scattered neutrons (Enderby & Nielson, 1979).

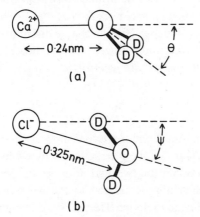

(a)

(b)

although the ranges over which r and θ can vary increase with an increase in the ionic radius (i.e. with a decrease in the surface charge density), as is to be expected. The angle θ is a function of the ion concentration, tending to zero at infinite dilution, and the number of water molecules in the primary hydration shell is six (four in the case of Li^+). The same arguments apply to the hydration of monatomic anions. Here, too, the hydration number is six, but the spatial localization becomes progressively less well defined as the ionic radius increases. The neutron diffraction studies indicate that the cation hydration shell is not penetrated to a significant degree by the anion, a result which is at odds with spectroscopic measurements of ion hydration (Hertz, 1973).

As was already stressed for the case of pure water, the structural studies provide no information about the residence time of a given water molecule at a site in the ionic hydration shell. Nuclear magnetic resonance (NMR) measurements provide such estimates. The indications are that some ions (Li^+, Mg^{2+}, Ca^{2+}, transition metals) reduce the rate of exchange of water compared to that in pure water, whereas other ions (K^+, Rb^+, Cs^+) enhance the exchange rate (Hertz, 1973). In other words, the residence time of a water molecule in the Ca^{2+} hydration shell is longer than that of a water molecule hydrogen bonded, on average, to four other water molecules in pure water. The effects produced by anions on the mobility of water are more pronounced than those due to cations, with SO_4^{2-} and PO_4^{3-} retarding the mobility of water and I^-, CNS^-, NO_3^- and ClO_4^- enhancing it.

There is as yet little information about the distributions of water molecules beyond the primary hydration shell. Since the electric field due to a monatomic ion is of a radial nature, the spatial and orientational dispositions of the O—H vectors of the water molecules in the primary hydration shell are incompatible with those in bulk water, so that there must arise a region of structural mismatch (Frank & Wen, 1957). The exact range and nature of this region will depend on the charge distribution and charge density of the ion. Operationally this effect is reflected in the equation relating the viscosity η of an electrolyte solution to the concentration:

$$(\eta - \eta_0)/\eta_0 = Ac^{\frac{1}{2}} + Bc + \ldots \tag{1.1}$$

where η_0 is the viscosity of water and A and B are constants (Jones & Dole, 1929). Qualitatively B is a measure of the interaction between hydrated ions and of the interference between the two hydration shells. The signs and magnitudes of the B-coefficients are related to the water mobility enhancement or retardation referred to above (Hertz, 1973). The observed effects are frequently termed '*structure making*' and '*structure breaking*', respectively, but such descriptions are misleading, because they leave the reference state of 'structure' vague. Certainly, in the immediate neighbourhood of a Li^+ ion, the positions and orientations of water molecules are much better defined than they are in bulk water. They are, however, completely different from those that exist in bulk water, so that the degree of structure in the ionic hydration shell has been enhanced, but the kind of structure is different from, and incompatible with, the characteristic tetrahedral arrangement in liquid water.

Turning now to the hydration behaviour of nonionic solutes, it is possible to distinguish two types of molecules (Franks & Reid, 1973): those which are apolar or predominantly apolar, with only one functional group, and those which are predominantly polar with two or more functional groups capable of participating in hydrogen bonding. Examples of the first type, apart from the hydrocarbons themselves, are homologous series of alkyl derivatives: alcohols, ethers, ketones and amines, while examples of the latter type are sugars, sugar alcohols, polyamines and amides. The interactions of water with the first group are of great importance to biochemists because they provide a large part of the forces that hold biological structures together (Kauzmann, 1959).

The introduction of an apolar residue (or molecule) into water results in a rearrangement of the water molecules, so that a cavity is created which is capable of accommodating the chemically inert residue. The important feature of this rearrangement is that it takes place without a decrease in the water–water hydrogen bonding or in the hydrogen bond energy

Fig. 1.3. Typical concave water molecule cage formed around a nonpolar solute particle. The particular configuration shown involves 14 water molecules of which 8 are close neighbours of the solute and 6 are placed at a somewhat greater distance. The configuration is governed by the tetrahedral hydrogen bonding pattern which must be maintained.

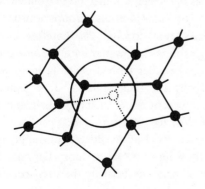

(Geiger, Rahman & Stillinger, 1979; Rossky & Zichi, 1982). A typical water molecule arrangement is shown in Fig. 1.3. One important consequence of the reorganization of water molecules round the apolar residue is a reduction in the permitted configurations available to the water molecules, since O—H groups must not be directed towards the centre of the cavity; they may point along the edges or away from the cavity. The phenomenon, referred to as hydrophobic hydration, is thus of a configurational origin and takes place with a loss of degrees of freedom, equivalent to a decrease in the entropy of the water (Franks, 1975). Dynamic measurements indicate that the water molecules which are involved in forming the hydration shell suffer a retardation in their diffusive motions, while the solute molecule, on the other hand, is free to rotate rapidly within the hydration cavity (Goldammer & Zeidler, 1969; Leiter, Patil & Hertz, 1983). As was the case for the ionic hydration shell, little is yet known about the range of the hydrophobic hydration effect, although there is a growing opinion that it extends over several water molecular diameters (Pratt & Chandler, 1977; Pashley & Israelachvili, 1981). Since the origin of hydrophobic hydration is in the nature of a structural repulsion, any process that can counteract this reorganization and partial immobilization of water must be spontaneous (in the thermo-dynamic sense). Thus, two or more apolar residues, each with its associated hydration shell, can interact and form a joint hydration shell with an economy in the number of water molecules required. While some water molecules may then suffer even greater restrictions on their configurational and dynamic freedom, a proportion of water molecules will now be able to relax to their 'normal' bulk water state. The net result is an aggregation of apolar residues; that is, the aggregation is accompanied by a net decrease in the free energy. However, it must be strongly emphasized that this molecular association is not the result of an attraction between the apolar groups but the consequence of the structurally unfavourable reorganization of water molecules in the neighbourhood of isolated apolar groups, and the partial relaxation of water as apolar groups approach one another (Franks, 1975). An important corollary is therefore that the (negative) free energy of aggregation is dominated by a *positive* entropy term which becomes progressively smaller with decreasing temperature. Any biological structures that rely for their stability on a significant contribution from hydrophobic interactions are therefore likely to become more labile, or even unstable, at low temperatures.

Not much is known about the hydration details of highly polar molecules, especially those with several —OH groups (Suggett, 1975; Franks, 1979). This is unfortunate, because these are the very species that

are implicated in natural cold hardening processes. Presumably carbohydrate derivatives interact with water primarily by hydrogen bonding.
The hydrogen bond energies are likely to be very similar to those between
water molecules, so that hydration geometries and energies are likely to
be critically dependent on the spatial dispositions of —OH groups relative
to those that exist in bulk water. This, in turn, implies that the stereochemical details of the solute molecule determine the nature and stability
of the hydration shell. Figure 1.4 illustrates how minor stereochemical
changes (in this case the out-of-plane flexing of a furanoid ring) can produce quite significant alterations in the —OH spacings and orientations
at the periphery of the molecule. Such alterations affect the nature and
range of the hydration shell.

A very sensitive indicator of hydration is provided by the solute partial
molar heat capacity, measured at infinite dilution, $\overline{C}_p^0$. It is a measure of
the thermal lability of the hydration shell. Table 1.1 summarizes some data
for some molecules of similar sizes and shapes. A comparison with similar
data for ions is instructive. Dealing with the ions first, electrostatic theory
predicts that $\overline{C}_p^0 < 0$ and that its magnitude is proportional to the
inverse ionic radius (Friedman & Krishnan, 1973). The experimental values
for the cations show that $\overline{C}_p^0$ becomes increasingly negative (less positive)
with increasing ionic radius, a result which cannot be accounted for in

Fig. 1.4. The effect of furanose ring flexibility on the peripheral —OH
distributions (only oxygen atoms are shown) in two sucrose-containing
trisaccharides. The inversion of the positions of two carbon atoms in
the ring drastically changes the mutual spacings of the substituent
—OH groups. After Jeffrey (1973).

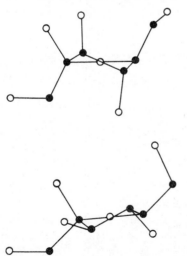

Table 1.1. *Limiting partial molal heat capacities of molecules and ions in aqueous solution at 25 °C*

Solute	Mannitol	Sorbitol	Inositol	Glucose	Galactose	Fructose	Benzene
Molecular mass	182	182	180	180	180	180	78
$\bar{C}_p^0$ J (K mol)$^{-1}$	451	410	340	347	324	352	360

Solute	Li^+	Na^+	K^+	Rb^+	Cs^+	Mg^{2+}	Ca^{2+}	Bu_4N^+
$\bar{C}_p^0$ J (K mol)$^{-1}$	176	155	130	125	109	259	226	1467

Solute	F^-	Cl^-	Br^-	I^-	OH^-	NO_3^-	ClO_4^-	SO_4^{2-}	PO_4^{3-}	OAc^-
$\bar{C}_p^0$ J (K mol)$^{-1}$	238	247	255	260	251	196	167	481	945	113

terms of only the primary hydration shell. A similar trend is observed for the monatomic anions. As regards the polyatomic ions, some allowance should be made for the contribution to $\overline{C}_p^0$ from the internal degrees of freedom, but even when this is done, the $\overline{C}_p^0$ values are incompatible with a simple electrostatic picture of an ion–dipole effect. Although at present the results cannot be interpreted quantitatively, they strongly suggest the existence of perturbed water at some distance from the ion. The structural and energetic details of such perturbed water depend sensitively on the ion charge density and charge distribution, especially in the case of polyatomic ions such as the oxyanions. The degree of compatibility of the secondary hydration shells of different ions determines the physical behaviour of concentrated solutions, as expressed for instance through the viscosity B-coefficients in eqn (1.1).

The interpretation of $\overline{C}_p^0$ data of nonelectrolytes is even less clear cut, because there are no molecular theories to provide a reference basis. For homologous series of n-alkyl derivatives $\overline{C}_p^0$ increases by a constant amount (approx. 85 J (mol K)$^{-1}$) for each additional —CH_2 group (Konicek & Wadsö, 1971). This is considerably in excess of the contribution due to the additional vibrational and rotational degrees of freedom. This constancy per —CH_2 addition is consistent with the hypothesis of water cage stabilization by apolar residues. A comparison of substantially apolar molecules with polyhydroxy compounds having the same number of carbon atoms shows the latter to have much smaller $\overline{C}_p^0$ values; compare, for instance n-hexanol with mannitol or sorbitol. Despite the larger size and greater number of chemical bonds, the polyhydroxy compounds have substantially lower $\overline{C}_p^0$ values, indicating that their interactions with water produce less thermolabile structures than do those of water with apolar substances. The observed results can also be taken as indications of shorter range hydration interactions.

More subtle differences in the hydration behaviour become apparent from a comparison of stereoisomers, e.g. xylitol, arabinitol and ribitol, or glucose, galactose and mannose (DiPaola & Belleau, 1977; Lian, Chen, Suurkuusk & Wadsö, 1982). There is, however, not enough known about the interactions between water and polyhydroxy compounds for such differences to be interpreted.

1.4. Hydration, as reflected in solute–solute interactions

Of greater practical significance than the interactions of water with isolated molecules (infinite dilution) are the interactions between solute molecules in an aqueous medium, or the interactions between functional groups within a macromolecule. In terms of thermodynamic representation, such effects are conveniently expressed in terms of excess functions, defined as

$$\Delta X^{e} = \Delta X_{\text{expt}} - \Delta X_{\text{ideal}}$$

where X is any function of state. ΔX^{e} can be expressed in terms of a power series of the type

$$\Delta X^{e} = x_{ii}c_i^2 + x_{iii}c_i^3 + \dots \quad (1.2)$$

where c_i is the concentration of species i, and x_{ii} etc. are fitted virial coefficients which can be related to the interaction between two, three, or more molecules (Kozak, Knight & Kauzmann, 1968). When $X = G$, the Gibbs free energy, then g_{ii} is related to the second virial coefficient in the osmotic pressure equation which relates a_w to the solute concentration. A net attraction between two molecules in solution gives rise to $g_{ii} < 0$ and a net repulsion to $g_{ii} > 0$.[†] Equation (1.2) can be extended to ternary solutions, when the interactions between different species i and j will be characterized by a coefficient x_{ij}. Table 1.2 summarizes some second virial coefficients for binary and ternary systems, obtained from measurements of the concentration dependence of vapour pressure, enthalpy, density and specific heat. The selected data in Table 1.2 permit certain conclusions to be drawn: (1) hydrophobic molecules *appear* to attract each other, but this is not due to an exothermic mixing effect but to the *net* favourable entropy

Table 1.2. *Second virial coefficients x_{ii}, x_{jj} and x_{ij} where x is free energy (g), enthalpy (h) or entropy (s), see eqn* (1.2).

Data are obtained from dilution and mixing experiments and refer to 25 °C. Units: J mol^{-1}(mol kg^{-1})$^{-1}$

i	j	g	h	Ts
Ethane	Ethanol	−200	—	—
Ethanol	Ethanol	−125	243	368
Tetrahydrofuran	Tetrahydrofuran	−207	1182	1389
Urea	Tetrahydrofuran	−17	295	312
Urea	Urea	−106	−351	−245
Inositol	Inositol	−260	−800	−540
Glycerol	Glycerol	15	362	347
Mannitol	Mannitol	8	66	58
Sucrose	Sucrose	174	577	403
Sucrose	Mannitol	91	—	—

[†] Strictly speaking, the Helmholtz free energy ($A = G - PV$) provides more unambiguous information about molecular interaction, because A is measured at constant volume. In practice, however, G (at constant pressure) is easier to measure experimentally. It is usually assumed that the PV term is negligible, an assumption which can give rise to significant errors with some aqueous solutions.

of the hydrophobic interaction. The magnitude of the hydrophobic interaction increases with the size of the alkyl group but is also influenced by its shape. (2) Polyhydroxy compounds appear to exhibit net repulsions (or at least very weak attractions), with the exception of inositol which resembles urea in the relative magnitudes of the x_{ii} coefficients. The enthalpy of the molecular pair interaction (h_{ii}) is positive for small molecules (endothermic mixing), becoming negative with increasing molecular weight. (3) Within a group of stereoisomers (e.g. the aldohexa-pyranoses or the hexitols) there are significant variations in h_{ii} (and probably in g_{ii}). (4) The interactions between unlike molecular species cannot be predicted from the corresponding interactions between like molecules; h_{ij} is not usually the mean of h_{ii} and h_{jj}. (5) The molecular pair interaction coefficients do not bear an obvious relationship to the hydration interactions, i.e. those measured at infinite dilution.

The role played by hydration in modifying solute–solute interactions cannot be extracted from the ΔX^{e}, as given by eqn (1.2), because the theory on which equations of the type (1.2) are based assumes that the solvent can be regarded as a continuum and that specific distributions and orientations of solvent molecules need not be taken into account (McMillan & Mayer, 1945). Nevertheless, the subtle differences in the x_{ij} coefficients can only be accounted for on the basis of interference – constructive or destructive – between molecular or ionic hydration shells (Lilley, 1973). This is shown diagrammatically in Fig. 1.5. Molecular interactions in solution are thus seen to be long range effects, modulated by the hydration

Fig. 1.5. Solute–solvent interaction in solution: A – primary hydration shells; B – secondary hydration shells; D – outer shell overlap; E – primary shell overlap. After Lilley (1973).

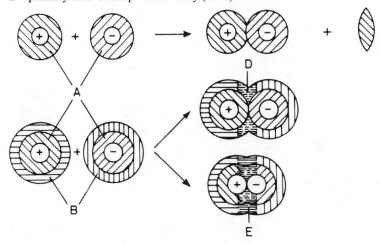

shells of the two interaction partners. At the present time there are no theories which would make possible a detailed specification of these hydration shells. On the other hand, any theory which purports to account for interactions between molecules without consideration of the molecular nature of the solvent is suspect.

The influence of molecular interactions on hydration and *vice versa* is of particular importance in processes involving macromolecules, especially biomacromolecules. Consider the peptide chain of a small globular protein: in its folded state there is a multiplicity of group interactions involving both the backbone peptide groups and the side chain residues. On the periphery of the macromolecule these various functional groups interact with the aqueous solvent, and through the solvent, with one another, very much in the manner outlined above. However, even in the interior of the folded molecule water molecules are not excluded. Indeed, neutron diffraction studies have identified many water molecule sites integral to the protein molecular structure in the crystals (Finney, 1979). Several examples of such sites of water molecules are shown in Fig. 1.6. Water is thus seen to play a part in the maintenance of the native conformation, by providing hydrogen bond links between backbone carbonyl and —NH groups and also by hydrogen bonding to certain side chains. Water molecules are also situated close to charged groups in the interior of globular proteins, where they reduce electrostatic repulsion between such groups. It will be shown presently that the stability of proteins towards changes in external conditions depends very much on their ability to maintain a minimum degree of hydration, and that physiological and biochemical safeguards are designed to achieve this

Fig. 1.6. Locations of water molecules (large circles) in protein crystals; hydrogen bonds are shown by broken lines. *a*: carbonic anhydrase-B; *b*: papain and *c*: pancreatic trypsin inhibitor. After Finney (1979).

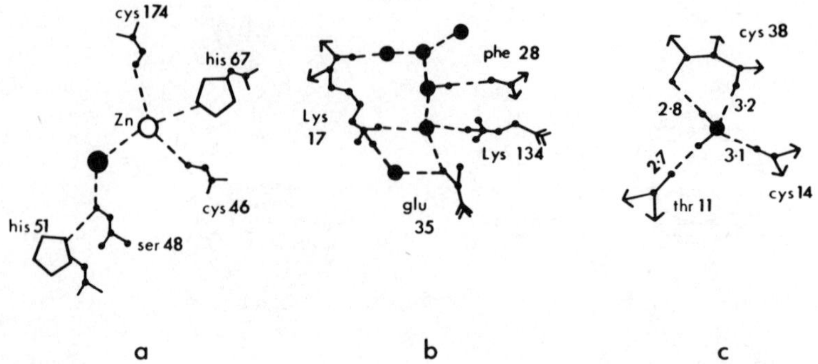

under extreme stress conditions. Such protection mechanisms are extremely subtle. To express such effects simply in terms of a_w is quite unrealistic, as will be shown in more detail in Chapters 3 and 4.

The above discussion serves to demonstrate the complexity and subtlety of molecular interactions in aqueous media because of the diverse ways in which different functional groups interact with water and one another. Where different functional groups exist in close proximity, such as in a single molecule, say of an amino acid, or where they exist as side chains on a common polymer backbone, as in a protein, the contribution of the various hydration effects can only be expressed as an averaged contribution to the overall free energy.[†] Nevertheless, the realization that such different hydration interactions are involved is helpful in the elucidation of temperature and concentration effects on molecular stability and interactions.

1.5   Life, low temperatures and freezing

It must surely be a coincidence that the freezing point of the liquid which is essential for life lies almost exactly at the centre of the temperature range which we associate with life on this planet. Subzero temperatures are very common in the terrestrial environment, and in order to survive, living organisms have developed the means for coping with the twin effects of low temperature and freezing. As was pointed out above, the term 'low temperature' is a relative one; on the other hand, freezing is exactly defined as a first order phase transition which involves the crystallization of liquid water. If this water crystallizes from an aqueous solution, then the process of freezing is accompanied by a gradual increase in the concentration of all soluble species in the residual liquid phase. This freeze concentration effect can have disastrous consequences on the viability of cells.

Although the equilibrium freezing point is exactly defined as the temperature at which the free energy curves of the solid and liquid states intersect, the phenomenon of undercooling is common and well documented. By cooling aqueous solutions under conditions where ice will not readily separate at the equilibrium freezing point, the injurious effects

---

[†] Nevertheless, various schemes have been proposed according to which the overall interaction energies can be allocated, on an additive basis, to individual groups of atoms, e.g. CH_3, CH_2, CH_2OH, $>C=O$ etc. (Nichols *et al.*, 1976; Okamoto, Wood & Thompson, 1978; Blackburn, Lilley & Walmsley, 1980). While such group additivity schemes are moderately successful in accounting for, or predicting, free energies, they cannot account for differences in ΔH^e, ΔV^e or their higher T and P derivatives observed, for instance, within groups of stereoisomers such as xylitol, ribitol and arabinitol (DiPaola & Belleau, 1977). In any case such group additivity schemes are unsatisfactory from a theoretical point of view (Franks & Pedley, 1981).

of freeze concentration can be avoided. There is a limit to the degree of undercooling that can be achieved, and the lower is the temperature of the undercooled liquid, the more susceptible it becomes to chance freezing. Nevertheless, the principle of undercooling finds extensive application in low temperature survival mechanisms.

The physical properties of liquid water change dramatically and discontinuously at its freezing point. This is illustrated by the temperature dependence of the molar heat capacity which is shown in Fig. 1.7. At 273.2 K the heat capacity decreases by 50% if the water freezes. If, on the other hand, the water is able to undercool, then the $C_p(T)$ curve exhibits no discontinuity (Angell, 1982). Indeed, as the temperature decreases to below 260 K, C_p rises increasingly sharply with decrease in temperature, so that the heat capacity difference between ice and water increases correspondingly. Similar effects are observed for most other physical properties, e.g. density and compressibility (Angell, 1982). The origin of this rapidly increasing temperature sensitivity of the physical properties of undercooled water is still under active investigation, but it must be concluded that, at any given temperature, ice and undercooled water have dramatically different physical properties. Such differences are even more pronounced in the dynamic properties of dissolved ions and can amount to orders of magnitude, as shown in Table 1.3. When considering viability at subzero temperatures, it is therefore necessary to differentiate clearly between the effects produced by the low temperature as such, and those

Fig. 1.7. Heat capacities of liquid water and ice as functions of temperature. After Angell (1982).

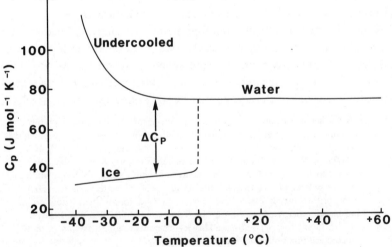

due to freezing (Franks, 1981*b*). Since life processes are so sensitive to the energy of the hydrogen bond as it exists in liquid water, and since the hydrogen bond energy is so sensitive to changes in temperature, it is hardly surprising that biochemical and physiological processes are affected by changes in temperature and the composition of the aqueous medium which, in purely chemical reactions, might be considered of minor impact.

The physico-chemical effects leading to low temperature water stress symptoms can be generalized as follows: when freezing is involved, the injury is due primarily to concentration effects. Where the low temperature is not accompanied by freezing, damage arises mainly from factors such as changes in ionic activities, reduction in diffusion rates and a disruption of the energetic balance which is responsible for the maintenance of biologically significant structures – membranes, enzymes and other macromolecular complexes, and differential changes in the rates of coupled reactions. In terms of physiology and ecology the two stress conditions are described as freezing injury and chill injury, respectively.

1.6   Strategies for survival

A survey of the manner in which different organisms cope with physiological water stress, whether it be induced by temperature or by salt concentration, indicates two major types of mechanisms: tolerance and resistance. They will be described in detail in Chapters 6 and 7. At this stage only the major features and differences will be considered. The two strategies can best be described in terms of their mechanical analogues. Tolerance to an applied stress corresponds to the phenomenon of viscosity or viscoelastic, plastic deformation in a solid. Here the solid absorbs the energy produced by the stress and is itself deformed, but is still able to function to some extent. The response time to the applied stress is important; it must be fast enough to absorb the stress. A solid body that has been subjected to a plastic deformation can never fully recover its initial dimensions. Plastic deformation is therefore irreversible. In

Table 1.3. *Dynamic properties of ions in water and in ice*

	Water	Ice
Proton mobility ($m^2 V^{-1} s^{-1}$)	36	3000
Li^+ mobility ($m^2 V^{-1} s^{-1}$)	4	$< 10^{-4}$
Li^+ self diffusion coefficient ($m^2 s^{-1}$)	0.24	8×10^{-7}

biological systems, on the other hand, the effects of the stress can be repaired by the expenditure of metabolic energy, even when the organism is inactive (dormant) during the application of the stress. Examples of this type of response are provided by freeze tolerance in plants and insects or tolerance to substantial desiccation by organisms exposed to dry heat.

/The phenomenon of resistance to an applied stress has its mechanical analogue in elastic deformation, where the body provides a restoring force in an attempt to reduce the stress. The danger of this type of response is that every material has an elastic limit, beyond which it breaks. Among living organisms stress resistance is encountered mainly through the phenomenon of undercooling which is particularly common among woody plants, subjected to prolonged periods of cold. Here, too, the threat of inadvertant freezing is ever present, since undercooled water is metastable with respect to ice and becomes increasingly so with decreasing temperature./ This is, in addition, a lower limit to the temperature at which freezing can be avoided. This limiting temperature depends on the chemical composition of the tissue fluids and on the manner in which the water is distributed throughout the organism.

In many cases survival is achieved by a combination of tolerance and resistance, so that a clear distinction between the two effects is not always realistic. However, the mechanisms and symptoms characteristic of the two survival strategies are quite distinct and can be studied in isolation by whole organism experiments or with the aid of isolated organelles, membrane or enzyme systems.

1.7   Low temperature preservation in medicine and industry

The long term storage of materials of organic origin and with a high water content presents many problems, not least of them their susceptibility to microbial deterioration. The inhibition of biochemical and microbiological spoilage of labile materials by storage at low temperatures has been practised for many years, but the degree of refinement of the methods employed has not kept pace with developments in our understanding of the fundamentals of freezing, crystallization and thawing. Thus, the overriding objective in the freezing of bulk food material is microbiological safety. While this may be quite understandable, it is nevertheless surprising how little attention appears to be paid to the preservation of the textural quality of the products. Indeed, many of the standard freezing and storage protocols are carried out under conditions which result in physical deterioration of the thawed product. A better appreciation of the factors responsible for microbiological safety and physical or chemical preservation is likely to lead to considerable improvements in the quality of freeze preserved products. Quite distinct from the application of freezing for purposes of storage, low temperature technology is also used by the food industry as a processing aid in the manufacture of products which are to be eaten in the frozen state (e.g. ice cream) or which are subsequently to be dried (freeze drying).

The medical applications of low temperatures date from the discovery (actually a rediscovery) in 1949 that glycerol can protect living cells against injury during freezing (Polge, Smith & Parkes, 1949). The potential benefits of cryopreservation to medicine and agriculture are obvious, and a considerable effort has gone into the establishment of suitable preservation protocols for a wide variety of different cells, tissues, organelles, organs and even organisms (Ashwood-Smith & Farrant, 1980). In basic terms, the science of cryobiology has not advanced significantly during the past twenty years. It relies mainly on the identification of suitable cryoprotectants and the achievement of suitable cooling and warming rates for maximum recovery. Major successes have been achieved in the storage of blood components, a development which has revolutionized the technology of blood banking. More recently it has become possible to freeze mammalian embryos for long term storage and eventual transfer to the reproductive tract of a foster mother.

The possibility of storing organs and tissues is a *sine qua non* for any significant advances in transplantation, and scientists are hard at work in efforts to devise suitable methods for the long term storage of kidneys and other organs at liquid nitrogen temperatures. The practical problems are severe, and with present knowledge the chances of success cannot be rated very high.

The potential clinical applications of cryopreservation have stolen the limelight from another important area, agriculture, where the development of suitable methods might lead to much more revolutionary changes. Current trends in crop breeding, with the substitution of cultivars for wild varieties, lead to an ever more limited gene pool, with wild varieties becoming extinct. To allow this process to continue unchecked is unwise and dangerous in the extreme, bearing in mind future needs for food production. The preservation of plant genotypes, either in the form of cultured cells or meristem tissue, should not present problems as severe as those encountered in the preservation of kidneys or hearts. However, the effort devoted to the solution of such problems, and to the collection of plant varieties threatened with extinction, is almost nonexistent. That the establishment of cell and tissue banks is possible, even on an international scale, has been amply demonstrated by those industries which utilize microorganisms. Thus, the brewing industry disposes over extensive collections of yeasts, maintained in frozen or freeze dried storage.

For the sake of completeness two further aspects of low temperature in medicine must be mentioned: cryosurgery and problems relating to hypothermia and the injurious effects of frost. Very low temperatures are used as surgical tools mainly for three purposes: (1) as physical methods in the surgical handling of tissues, (2) for neurological inhibition and the destruction of tissue and (3) as specific treatments in cancer therapy. The study of hypothermia is closely related to that of hibernation and dormancy. Apart from accidental hypothermia, the condition can also be induced and is said to provide beneficial effects in the treatment of some diseases.

The enormous potential of cryopreservation and long term storage has led to some strange speculations and, even worse, to some commercial excesses. They involve the freezing of human cadavers for indefinite periods, on the assumption that they can be revived at some future date. Reports of such activities have appeared in the press since 1968, and companies were formed for the express purpose of freezing and storing human bodies for future resuscitation (Synon, 1979). Such excesses of cryobiology can only serve to bring the scientific discipline into disrepute. That is not to deny that at some future date it may become possible to preserve whole organisms in a state of suspended animation. However, to resuscitate a human cadaver after brain death has occurred presents problems of a much greater order of magnitude.

I have touched on this aberration of science and technology to emphasize the enormous potential of low temperatures for the almost unlimited preservation of labile materials which under normal circumstances have a very limited life span. Those organisms that have to exist under harsh climatic conditions can acclimatize to the stresses of low temperatures. Other types of cells, including those of mammalian origin, have no such acclimatization mechanisms, but with the aid of chemical protectants and special cooling and warming procedures, this lack of natural protection may eventually be overcome and long term survival attained. As of now, only fairly small groups of cells, or fairly simple tissues, can be successfully preserved, but there appear to be no fundamental reasons why such successes could not be translated to whole organs or even organisms, just as it should, in time, become possible to develop organisms (especially plants) that possess an enhanced degree of cold tolerance. The economic and social implications of such developments have already been the subject of some discussion and controversy, but the technical problems should be no more severe than those which have from time to time delayed other major scientific, medical or technological innovations.

2

The physics of water at subzero temperatures

2.1   Terminology

The peculiar structure of the water molecule, illustrated in Fig. 1.1, gives rise to extensively hydrogen bonded networks in which a tetrahedral coordination predominates; this is true even for the liquid state. In other words, each oxygen atom is surrounded on an average by four other oxygen atoms, placed at, or close to, the vertices of a regular tetrahedron (as shown in Fig. 2.1) with a hydrogen atom placed along each O—O axis.

Fig. 2.1. Crystal structure of 'ordinary' ice (Ih), showing the regular hexagonal lattice of oxygen atoms (dark circles) with one hydrogen atom (light circle) lying on each O—O axis. Each oxygen atom is thus hydrogen bonded to four other oxygen atoms, placed at the vertices of a regular tetrahedron.

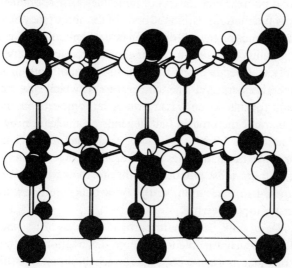

In the case of the solid, ice, this arrangement is regular with O—O distances of 0.275 nm and H—O—H angles corresponding to those of a regular tetrahedron, 109° 28′. In liquid water these distances and angles suffer a considerable distortion. However, detailed X-ray and neutron scattering studies strongly suggest that the essentially 4-coordinated molecular distribution, characteristic of ice, is maintained in the liquid (Narten & Levy, 1972; Dore, 1981). The molecules in the liquid are of course subject to rapid diffusional motion, so that the term 'structure', as applied to a liquid, describes a time-averaged distribution of molecules.

Unlike a close packed structure, in which each (spherical) molecule is surrounded by twelve nearest neighbours, the 4-coordinated assembly of water molecules contains a large amount of empty space (approx. 80%). This, coupled with the fact that hydrogen bonds are labile, makes such a structure very susceptible to changes in temperature and pressure. The pressure/temperature phase diagram of ice bears witness to this.

The equilibrium freezing point of water is 0 °C (273.16 K); it is the temperature at which liquid water and ice coexist in equilibrium at atmospheric pressure and is denoted by T_f^0. In practice, when liquid water is cooled, it does not spontaneously freeze at T_f^0, but at some lower temperature. Undercooling is a common feature of liquids and, since it is of great relevance to the phenomenon of biological freeze resistance, it will be further discussed below. The crystallization of a solid phase from a liquid (or a vapour) requires the generation of nuclei onto which molecules can condense. An effective nucleus consists of a group of molecules which can be 'recognized' by other, diffusing molecules as a structure resembling that of the solid phase. Such clusters of molecules arise spontaneously by random density fluctuations within the body of the liquid; they have a finite lifetime which depends on the self-diffusion rate; eventually they dissociate. Thus, the probability that such a cluster can serve as an effective nucleus for crystallization depends on its size and its lifetime; both of these are functions of the temperature. The temperature at which this probability of condensation approaches unity is known as the homogeneous nucleation temperature, T_h, and the process that depends on nucleation by random density fluctuations is known as *homogeneous nucleation* (Turnbull & Fisher, 1949). Once nucleation has occurred, then each nucleus will grow into a crystal. Under most practical conditions the crystallization of ice is rapid compared to nucleation, so that in a bulk liquid few nuclei are needed to ensure rapid freezing.

Most liquids in bulk contain impurities, such as dust particles or microcrystallites, which are able to facilitate the creation of a nucleus by the sorption of water molecules onto the surface of the solid particle. Such impurities can therefore be regarded as catalysts for nucleation, just as metal surfaces act as catalysts for gas reactions. The process whereby nuclei form on some foreign surface is known as heterogeneous nucleation and is characterized by a temperature T_{het}, where $T_{het} > T_h$. The indications are that T_{het} depends on the surface characteristics of the catalyst and on the dimensions of the catalyst particle (Fletcher, 1970).

If ice is made to crystallize under conditions of minimum undercooling (e.g. by seeding), large, well-formed dendritic crystals result. If, on the other hand, liquid water is allowed to undercool to a substantial degree, the resulting crystals are small. During rewarming, these small crystals gradually disappeared to yield larger crystals, a process known as recrystallization or ripening. An example of ripening is the gradually coarsening texture of ice cream when it is stored for prolonged periods in a domestic freezer.

2.2 Phase behaviour of ice

Ice can exist in eight stable, well defined crystal modifications (Hobbs, 1974); their respective ranges of stability are illustrated in Fig. 2.2 and the temperatures and pressures which define the coexistence of any three phases (triple points) are summarized in Table 2.1. Ice IV does not really exist; it is something of a historical curiosity. At one time it was believed to be a stable polymorph, but later work showed it to be a metastable form of ice V.

In addition to the stable varieties there exists a metastable low pressure polymorph, ice Ic (cubic), which can, however, not be obtained directly from either liquid water or from any of the other stable modifications. It is obtained when water vapour is allowed to condense on a cold surface to form so-called amorphous ice (Sceats & Rice, 1982). This is a non-crystalline form of solid water which, on being heated to above $-120\,°C$, is transformed first into ice Ic and then into the stable form ice Ih (hexagonal).

Table 2.1 shows that not all the crystalline forms of ice can coexist with liquid water; some of them, ices II, VIII and IX, are only stable in the presence of some of the other crystalline forms. It is interesting to note the pressure sensitivity of the freezing point of water. Thus, at pressures of 2200 MPa, water freezes above 82 °C. The abnormal pressure dependence of the freezing point is confined to ice I, i.e. T_f^0 decreases with applied pressure. This is a symptom of the exceptionally low density of this polymorph; all the other crystal forms are denser than liquid water.

All the known crystal structures have a common feature: every oxygen atom is linked by hydrogen bonds to four other oxygen atoms. It seems

Table 2.1. *Triple point data for liquid (L) water and ice polymorphs*

Coexisting phases	Pressure (MPa)	Temperature (K)
I–L–III	207	251.2
I–II–III	213	238.5
II–III–V	344	248.9
III–L–V	346	256.2
V–L–VI	626	273.4
VI–L–VII	2200	354.8
VI–VII–VIII	2100	273

Fig. 2.2. Solid–liquid phase diagram of H_2O. Broken lines represent approximate, and dotted lines estimated or extrapolated, phase boundaries.

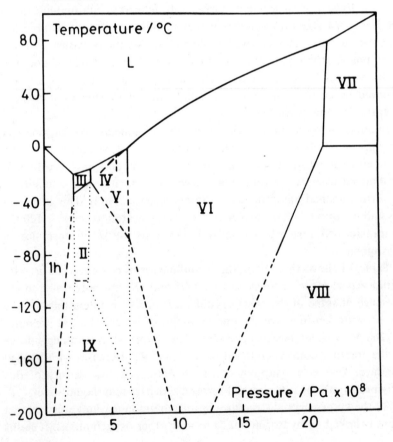

that pressure alone is not able to break hydrogen bonds, although extensive bond bending and distortion does of course occur with increasing pressure. In fact, the very high pressure polymorphs VII and VIII possess structures which closely resemble two ice I lattices pushed inside one another. The densities of these two forms are therefore almost double that of ice Ih, and each oxygen atom has eight nearest neighbours: it is hydrogen bonded to four of them with the other four belonging to the other, interpenetrating ice I lattice (Kamb & Prakash, 1974).

Despite the large amount of empty space in the ice I lattice, the solid is reluctant to incorporate solutes within the crystal structure, that is, it does not form solid solutions with chemically similar substances (except D_2O). There are two known exceptions: NH_4^+ and F^- ions, both of which can replace oxygen atoms in the ice crystal.

The mechanical properties of ice under different conditions of temperature and pressure are of the greatest importance in geological and glaciological studies and in problems posed by permafrost, as well as in ice engineering which is rapidly gaining in importance. In common with many crystalline materials, the mechanical properties of ice, such as elastic moduli and plastic flow, are determined by the concentration and nature of crystal defects, and these, in turn, depend on the manner in which the ice was crystallized and on its thermal history (Hobbs, 1974). The mechanical properties are probably not of great importance to the subject of this book, so the interested reader should consult one of the many specialized texts.

2.3 Nucleation of ice in pure water

Reference has already been made to the need for a nucleus to be generated before a daughter phase can crystallize in a liquid mother phase. Such nuclei arise within the body of the liquid through random density fluctuations, and the theory of homogeneous nucleation treats the energetics and kinetics of the growth and decay of such molecular clusters and their ability to act as nuclei for the crystallization process (Turnbull & Fisher, 1949; Dufour & Defay, 1963; Hobbs, 1974). The theory is currently under review, because of its several alleged shortcomings (Katz & Spaepen, 1978; Rasmussen, 1982; Franks, Mathias & Trafford, 1984). We shall limit ourselves to a summary of the basic features which have important implications for the natural phenomena of freeze avoidance and tolerance in living organisms.

Consider a small spherical volume v in an undercooled liquid at temperature T, corresponding to a degree of undercooling of $\Delta T = (T_f^0 - T)$. If the molecules which make up the spherical cluster are assumed to be in an ice-like configuration, then the free energy of condensation is given by

$$\Delta G_t = RT \ln (p_{ice}/p_{liquid}) \tag{2.1}$$

where p_{ice} and p_{liquid} are the sublimation pressure of ice and the vapour pressure of the undercooled liquid, respectively. Since at subzero temperatures ice is the stable phase, $p_{ice} < p_{liquid}$, so that $\Delta G_t < 0$. Actually ΔG_t is the free energy change that accompanies the transfer of a mol of water from the undercooled liquid to the interior of the solid-like cluster. Equation (2.1) can be approximated to

$$\Delta G_t = -\frac{T \Delta H_f \Delta T}{(T_f^0)^2}$$

where ΔH_f is the latent heat of crystallization. Opposing the growth of the cluster are the forces due to surface tension (σ) between the liquid and the cluster surface (assumed to be ice-like). For a cluster of radius r, the net free energy accompanying the liquid–solid condensation is

$$\Delta G_{l \to s} = 4\pi r^2 \sigma - (\tfrac{4}{3})\pi r^3 \Delta G_t \tag{2.2}$$

Fig. 2.3. The volume free energy of nucleation, $\Delta G_{l \to s}$, of water, as a function of the cluster radius at -40 °C, according to eqn. (2.2). The value for σ used in the calculations is that obtained from the nucleation data of Michelmore & Franks (1982), as analysed by Franks, Mathias & Trafford (1984). $\Delta G_{l \to s}^*$ is estimated at 0.5×10^{-18} J and $r^* = 1.85$ nm, corresponding to a critical nucleus of ~ 200 molecules. Note that r^* and ΔG^* do not coincide with the values shown in Table 2.2, which are based on rather older estimates of $\sigma(T)$ and $\Delta G^{\ddagger}(T)$, made long before good physical data for undercooled water became available.

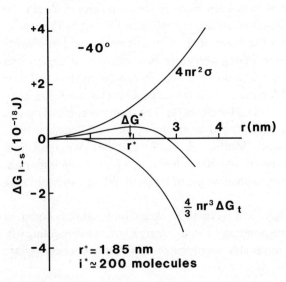

where σ is the interfacial free energy of the cluster. To estimate the value of r when $\Delta G_{1\to s}$ becomes negative, we differentiate eqn (2.2) with respect to r and equate the result to zero. This yields

$$r^* = -2\sigma/\Delta G_t \quad \text{and} \quad \Delta G^*_{1\to s} = 16\pi r^3/3(\Delta G_t)^2 \qquad (2.3)$$

as the critical values. The relationship between the various quantities in eqns (2.2) and (2.3) is represented in Fig. 2.3 for $T = 233$ K.

The number of critical clusters in a given volume of water at a given temperature is estimated by assuming a Boltzmann distribution:

$$n(r^*) = n_1 \exp(-\Delta G^*/kT) \qquad (2.4)$$

where n_1 is the number of molecules per unit volume in the liquid phase. It is now possible to express the volume of the nucleus in terms of the number of molecules it contains by assuming once again that the clusters are ice-like. The properties of water clusters and critical nuclei are computed in Table 2.2 as a function of temperature (Dufour & Defay, 1963). Consider a temperature of 243 K: $r^* = 1.38$ nm and i^*, the number of molecules in a critical cluster, is 566. The number of active nuclei in a gram of water is very small, well below the limit of experimental determination. From the data in Table 2.2 it is possible to calculate the mass of water which is likely to contain one critical nucleus at a given temperature; the results are shown in Fig. 2.4. From this we see that somewhere between 243 and 233 K a temperature is reached at which every gram of water is likely to contain one nucleus capable of initiating freezing. If the mass of water were now to be divided up into droplets of radius 10 nm, then, with the aid of the data in Table 2.2 and Fig. 2.4, it is possible to calculate the proportion of drops which are likely to contain a nucleus and will therefore freeze at that temperature. This is shown in Table 2.3 for several temperatures. The above discussion serves to demonstrate that nucleation is extremely sensitive to changes in temperature and that the

Table 2.2. *Dimensions and concentrations of critical nuclei in undercooled water*

Temperature (K)	Number of molecules per nucleus (i^*)	r^* (nm)	$\Delta G^*/kT$	Nuclei g^{-1}
263	15943	4.20	776	2.3×10^{-315}
253	1944	2.08	190	1.5×10^{-60}
243	566	1.38	83	3.8×10^{-14}
233	234	1.03	45	6.3×10^{2}
223	122	0.83	29	7.1×10^{9}

After Dufour & Defay, 1963.

Table 2.3. *Distribution of ice nuclei within dispersed water droplets (based on data in Table 2.2)*

Temperature (K)	Radius of drop (nm)	r^* (nm)	Droplets g^{-1}	Nuclei g^{-1}	Proportion of droplets containing a nucleus
263	10	4.9	2.4×10^{17}	2×10^{-433}	8.4×10^{-449}
253	10	2.26	2.4×10^{17}	8.2×10^{-75}	3.4×10^{-92}
	5	2.46	1.9×10^{18}	2.5×10^{-93}	1.3×10^{-111}
243	10	1.46	2.4×10^{17}	2.5×10^{-18}	1.1×10^{-35}
	5	1.55	1.9×10^{18}	1.4×10^{-23}	7.2×10^{-42}
233	10	1.08	2.4×10^{17}	6.3	2.6×10^{-17}
	5	1.14	1.9×10^{18}	0.029	1.5×10^{-20}

After Dufour & Defay, 1963.

Fig. 2.4. Mass of water (g) likely to contain one critical size nucleus as a function of temperature, assuming homogeneous nucleation.

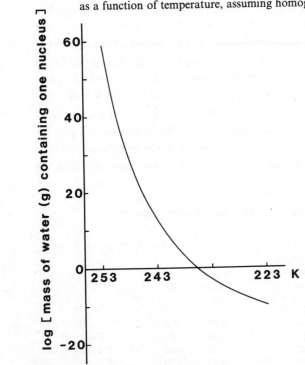

very large numbers involved in the calculations change rapidly by many orders of magnitude over small ranges of temperature.

Apart from estimating the size of critical clusters under varying conditions, it is also important to consider the kinetics of nucleation, that is, the rate of their generation in the body of the undercooled fluid. It is usually assumed that nucleation kinetics can be treated by transition state theory and J, the number of nuclei formed per second and per unit volume, is commonly written as

$$J \simeq \frac{n_1 kT}{h} \exp\left(\frac{-\Delta G^{\ddagger}}{kT}\right) \exp\left(\frac{-\Delta G^*}{kT}\right) \tag{2.5}$$

The second exponential term is the Boltzmann distribution of nuclei, as given by eqn (2.4) and the first exponential is the kinetic term which depends on the mechanism by which the clusters are formed. This is assumed to be by the stepwise addition of molecules diffusing to the cluster surface. $\Delta G^{\ddagger}$ is therefore taken to be the free energy of activation of self-diffusion. Since self-diffusion is related to fluidity (reciprocal viscosity), the activation free energy for viscous flow is commonly used in eqn (2.5). By insertion of the appropriate physical properties of water, we arrive at the following orders of magnitude for the various quantities in eqn (2.5) (Fletcher, 1970):

$$(n_1 kT)/h = 10^{41} \text{ m}^{-3} \text{ s}^{-1}$$

and $\quad [(n_1 kt)/h] \exp(-\Delta G^{\ddagger}/kT) \simeq 10^{36} \text{ m}^{-3} \text{ s}^{-1}$

in the neighbourhood of -40 °C. J should therefore increase rapidly with decreasing temperature. Near the nucleation threshold the rate changes by about a factor of 20 per degree, so that homogeneous nucleation is a well defined event which does not greatly depend on the rate of cooling.

By making various substitutons for ΔG^* in eqn (2.5), it is possible to express J in a fairly simple form:

$$J = A \exp(B\tau) \tag{2.6}$$

where τ is a quantity which contains all the temperature terms in eqn (2.5); it is given by $\tau = [T^3(\Delta T)^2]^{-1}$. A more useful way of expressing τ is by way of *reduced* temperatures, where $\theta = T/T_f$ and $\Delta\theta = (T_f - T)/T_f$ (Michelmore & Franks, 1982). This makes possible the comparison of different liquids and solutions which freeze at different temperatures. Equation (2.6) then becomes

$$J = A \exp(B\tau_\theta) \tag{2.6a}$$

where $\tau_\theta = [\theta^3(\Delta\theta)^2]^{-1}$. A plot of $\ln J$ against τ_θ should therefore be linear. The data in Fig. 2.5 confirm the validity of this relationship over several order of magnitude in J.

Having established that under optimum conditions, i.e. in the form of small droplets, water can be undercooled to approximately $-40\,°C$, corresponding to a reduced degree of undercooling $\Delta\theta = 0.14$, it is interesting to compare this with the abilities of other liquids to undercool. Table 2.4 provides such a comparison, showing water not to be particularly proficient at undercooling (Angell, 1982). The ability of water to undercool can be considerably enhanced by high pressures. This is illustrated in Fig. 2.6 which is an enlarged version of part of the ice phase diagram in Fig. 2.2.

Table 2.4. *Undercooling capacity of molecular liquids*

	T_f (K)	T_h (K)	$\Delta\theta$
Water	273.2	233.2	0.146
CH_3Br	179.4	155.0	0.16
Phosphorus	317.5	201.9	0.364
Ammonia	195.5	155.2	0.21
Benzene	278.4	208.2	0.252
Sulphur dioxide	197.6	164.6	0.20

After Angell, 1982.

Fig. 2.5. Temperature dependence of ice nucleation rate in undercooled water, plotted according to eqn. (2.6a), from DSC data obtained in the isothermal ($\bigcirc$) and scanning ($\bullet$) modes. After Michelmore & Franks (1982).

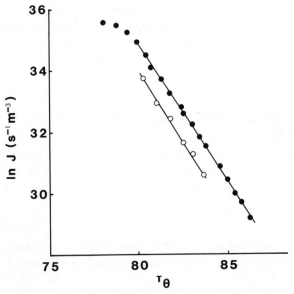

Starting at $-40\,°C$, corresponding to atmospheric pressure, T_h drops rapidly with increasing pressure, reaching $-90\,°C$ at a pressure corresponding to the ice I–ice III transition (Angell, 1982). It seems that for every degree depression of the freezing point, T_h decreases by approximately two degrees, although there is no obvious formal relationship between freezing and nucleation.

2.4   Nucleation of ice by particulate matter

In practice, the homogeneous nucleation of water or ice is a very rare event; it probably takes place in the upper atmosphere where supersaturated water vapour condenses to rain, snow or hail. Much more common is the process whereby nucleation is catalysed by a solid or liquid substrate which allows groups of sorbed water molecules to take up configurations that are able to promote further condensation. This process is known as heterogeneous nucleation, but it must be emphasized that the nuclei so produced are clusters of water molecules identical to those considered above. The function of the foreign surface is that of a catalyst: it enhances the probability that a cluster of critical dimensions can form.

Fig. 2.6. The effect of pressure on the homogeneous nucleation of ice in undercooled water, according to Angell (1982).

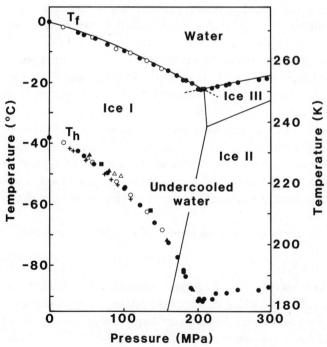

Much effort has gone into establishing what properties of the substrate are significant in rendering it active as a catalyst (Hobbs, 1974). Originally it was believed that a match of the crystal structure with that of ice was important, a hypothesis that resulted in the choice of silver iodide to promote nucleation of water and ice in supersaturated water vapour (Vonnegut, 1947). The nucleation thresholds for AgI, PbI$_2$ and CuS are -4, -6 and $-7\,^\circ$C respectively, showing these substances to be effective catalysts.

More recently it has been discovered that crystals of AgI doped with small amounts of AgBr form even more effective catalysts than pure AgI, indicating that lattice dislocations might be the catalytically active sites (Vonnegut & Chessin, 1971). Indeed, it has been suggested that the reason for the catalytic activity of AgI is that particles are subject to photolytic decomposition: free iodine and metallic silver are produced, the latter being oxidised in the presence of air. These impurity sites are thought to be responsible for the nucleating efficiency of AgI. The consensus is that three factors determine the nucleating efficiency of a substrate: (1) it should have a small lattice mismatch with ice, (2) it should have a low surface charge and (3) it should possess a degree of hydrophobicity, i.e. water should not spread on it.

In recent years attention has focussed on nucleating materials of biogenic origin. It is found that certain microorganisms, insects and plants contain materials which possess high concentrations of very efficient ice nucleators (Schnell & Vali, 1972; Krog, Zachariassen, Larsen & Smidsrød, 1979; Lindow, 1983). Indeed, as further elaborated in Chapter 6, heterogeneous nucleation is one of the mechanisms used by living organisms to minimize the undercooling of the body fluids. Studies in our laboratory have indicated that many, if not all, biological cells contain structures that are able to catalyse the nucleation of ice, although in some cases these structures are not of a high nucleating efficiency (Franks *et al.*, 1983). For instance, some structure (the plasma membrane?) contained in human erythrocytes promotes the freezing of intracellular water at a temperature which lies only 0.5 $^\circ$C above that of T_h of isotonic saline solution. Yeast cells, on the other hand, contain catalytic sites which promote freezing at 10° above T_h.

As regards the theory of heteronucleation, in general we can write

$$\Delta G^*_{\text{het}} = \Delta G^*_{\text{h}} f(m, R) \tag{2.7}$$

where R is a radius of the catalytic particle (assumed to be spherical) and m is a wetting parameter which describes the relative ease of wetting of the particle by ice and undercooled water respectively; $-1 < m < +1$ (Fletcher, 1970). For maximum catalytic efficiency ($T_{\text{het}} \simeq 0\,^\circ$C) $R > 10$ nm and $m \simeq 1$. Indeed, for a particle to possess any nucleating properties R must exceed 1 nm. The finer details of heteronucleation theory are uncertain, although various empirical or semi-empirical relationships have been established. For instance, for $R > 15$ μm, the undercooling temperature for a 50% probability of freezing is found to be

$$\ln(1/v) = aT + b$$

where a and b are constants (Bigg, 1953). What is certain is that, at any given temperature, $J_{het} \gg J_{hom}$. Thus, for heterogeneous nucleation the preexponential factor in eqn (2.6) is lower by many orders of magnitude than that corresponding to homogeneous nucleation.

Heteronucleation does not necessarily require the presence of foreign matter in the bulk or at the surface of the undercooled liquid. It can also be induced by a sudden compression, an electric field or irradiation. Indeed, the concept of nucleation was discovered by Fahrenheit in 1724, when he stumbled whilst carrying a flask filled with undercooled water and found that the water had frozen as a result of being shaken. The mechanism whereby extraneous factors can influence the nucleating potential of an undercooled liquid is completely unknown.

Before leaving the subject of undercooled water and the mechanism of ice nucleation, let us turn briefly to some of the features which raise doubts about the adequacy of the basic model, especially as applied to water; i.e. the stepwise growth of clusters until the critical dimensions for spontaneous crystal growth are reached. Mainly as a result of the comprehensive studies of the physical properties of undercooled water conducted by Angell and his associates (Angell, 1982, 1983), it now appears that the liquid becomes physically unstable in the neighbourhood of 228 K. Although nucleation intervenes at 233 K, many of the dynamic and thermodynamic properties show evidence of a divergence; this is illustrated in Fig. 1.7 for the heat capacity. Such behaviour is characteristic of a strongly cooperative process. The divergence of the physical properties of undercooled water can be completely suppressed in the presence of solutes, even those that chemically resemble water, e.g. H_2O_2.

A better and more realistic model for ice nucleation in undercooled water might need to be based on a cooperative growth of clusters, involving an increasingly large number of molecules with decreasing temperature. Evidence in favour of such a model is provided by X-ray scattering results on undercooled water (Bosio, Teixeira & Stanley, 1981). Quite apart from the divergence phenomenon, however, it might be questioned whether the stepwise growth model for nucleation is applicable to a liquid which already exists in the form of an infinite hydrogen bonded network and in which nuclei would presumably be formed by domains in which random density fluctuations result in hydrogen bond configurations which closely resemble those in ice. This, too, would need to be a cooperative process, involving more than one H_2O molecule at a time.

2.5   Ice crystal growth

Once nuclei of the critical dimensions exist in the undercooled liquid, growth of crystals will begin spontaneously. The crystal habits of ice are extremely complex (Hobbs, 1974), so that under experimental conditions which differ only marginally, needles, feathers, spherulites or disc type crystals can be produced. Eventually they all tend to be converted to the stable form, the familiar stellar dendrites which are formed directly by the condensation and solidification of water vapour. Crystal growth rates and morphologies depend critically on the degree of undercooling and the rate of cooling. Thus, we distinguish between equilibrium freezing which takes place under conditions of $\Delta T \simeq 0$, possibly with seeding, and perturbed freezing which is observed for $\Delta T \gg 0$. In the case of equilibrium freezing the undercooling of the water/air interface is of importance. An example is provided by the growth of ice on the surface of a pond which is cooled by radiation to the night air. In this case the latent heat is withdrawn through the ice, resulting in a smooth growth of the water/ice interface downwards. If ice grows within undercooled water, then the heat flows from the crystal to the liquid. This results in unstable crystals, because any one part of a crystal interface that gets ahead of its neighbouring regions can more easily dissipate latent heat and, in turn, grow faster. This type of freezing therefore leads to dendritic (tree-like) ice crystals. The rate of crystal growth is a function of the self-diffusion coefficient, whereas the concentration of crystals depends on the concentration of nuclei which is itself a function of ΔT.

Dendrites with their growing axis parallel to the thermal gradient predominate, and larger dendrites grow faster than small ones. Figure 2.7 shows the effect of the thermal profile on the ice morphology (Calvelo, 1981). The region shown as d is undercooled, so that nucleation is a likely event; it is most likely closest to the low temperature sink where the temperature corresponds to T_{het}. Nucleation causes the temperature to rise to T_f, resulting in the growth of dendrites which extend laterally and eventually touch, but they also grow in the direction of the temperature gradient. Usually only about 6% of the water will freeze in this manner, because the rising temperature inhibits further nucleation. Subsequent crystallization takes place by consolidation and conduction through the ice already present. At distances $> d$ no further thermal undercooling

Fig. 2.7. Thermal profile and ice morphology during freezing of water in a solid matrix; for explanation see text. After Calvelo (1981).

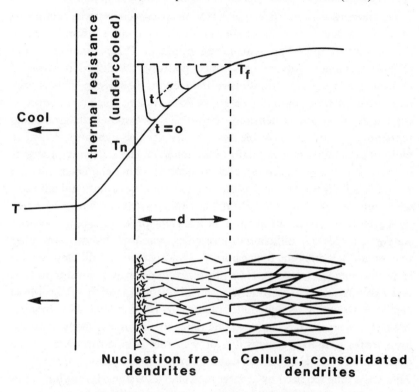

Nucleation free dendrites   **Cellular, consolidated dendrites**

Fig. 2.8. Dynamics of water molecules in heterogeneous systems, e.g. pores or capillaries. A number of characteristic diffusion and exchange times can be distinguished, some or all of which can be determined experimentally, depending on the system under study and the experimental conditions.

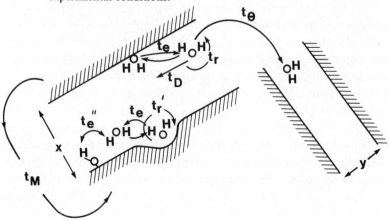

occurs and ice crystallization then takes place in the form of cellular dendrites.

The freezing of water from aqueous solutions, especially under conditions of rapid cooling, is even more complex, because concentration gradients are then superimposed on temperature gradients. This is discussed in the following chapter. Any process which perturbs the diffusion of water molecules is likely also to influence the freezing behaviour. This is very noticeable for water in highly heterogeneous systems, such as in pores or capillaries of $\leqslant \mu$m dimensions. Figure 2.8 provides a diagrammatic representation of the possible restrictions on the dynamics of water molecules confined in a capillary of diameter x. The following processes need to be taken into account, each associated with a characteristic time t: rotational diffusion (t_r), translational diffusion (t_D), rotational diffusion when interacting with the wall of the capillary (t'_r), exchange of water molecules between the surface and the bulk (t'_e), proton exchange between water molecules (t_e), diffusion of molecules from one channel to another one which has different dimensions (y) and is oriented differently with respect to some reference axis (t_θ), proton exchange between water protons and protons on the substrate (t''_e) and diffusional motion of the whole capillary, including its aqueous contents (t_M) (Packer, 1977; Derbyshire, 1982). Of these various characteristic times the last one is likely to be the longest; for globular proteins it is of the order of microseconds. At the other end of the scale, for bulk water at ordinary temperatures, $t_r = 3$ ps. This is the time required by a water molecule to perform a rotation about its axis. The effect of temperature on t_r is quite marked. Thus, near the homogeneous nucleation temperature ($-40\,°$C), $t = 30$ ps. Other diffusion and exchange times that have been reported include $t'_r = 1$ ns for water close to the surface of a globular protein at $-15\,°$C, $t_e = 7\,\mu$s and $t'_r = 23$ ns for water in muscle tissue (Derbyshire, 1982). The degree of dynamic perturbation induced by the proximity of surfaces also affects the freezing behaviour, to the extent that the water in very small pores or narrow capillaries may not freeze at all. However, even when some does freeze, a proportion always remains unfrozen. The incidence and importance of so-called unfreezable water will be touched upon in several of the following chapters.

Very recently Raman studies of ice at 100 K and pressures of up to 50 G Pa have produced evidence for a new polymorph, ice-X, in which the hydrogen bond is centrosymmetric and the O—H—O distance is 0.232 nm, the shortest hydrogen bond ever reported for water (Hirsch & Holzapfel, 1984).

3

Physical chemistry of aqueous solutions at subzero temperatures

3.1 Homogeneous solutions at low temperatures

Temperature is the manifestation of kinetic energy which is itself a measure of molecular motion. Changes in temperature are therefore likely to influence most properties of a molecular system. It has already been emphasized that the effects produced by changes in temperature are in no way related to those which accompany freezing, and that the very term 'low temperature' is a relative one. With aqueous systems a low temperature is often equated with a subzero temperature, but this is purely subjective. The existence and relative stability of undercooled water demonstrates that aqueous systems can persist as homogeneous mixtures well below their equilibrium freezing points. Here we are concerned with changes in physical properties of *homogeneous* aqueous systems at subambient temperatures. In the absence of freezing or the precipitation of one of the solute components, any given physical property of an aqueous solution is a continuous function of temperature. This was shown in Fig. 1.7 for the heat capacity of water. A survey of the physical properties of water shows that they respond very differently to temperature changes (Franks, 1981b). Some changes are monotonic (refractive index, dielectric permittivity) while others undergo maxima or minima (density, compressibility). Some changes are very marked (pH), while others are quite small (hydrogen bond length).

The influence of temperature on the ionization of water merits special attention, because it, in turn, determines the changes in acid and base ionization, a factor which must be taken into account, for instance, in the preparation of buffer mixtures (Taylor, 1981). It is a widely held misconception that pH = 7 denotes the universal condition of neutrality; in fact this is true only at 25 °C, where the hydrogen and hydroxyl ion activities happen to be 10^{-7} g ion dm^{-3}. However, in common with most

37

other homogeneous equilibria, the ionization of water is subject to changes with changes in temperature. The most reliable experimental data lead to the relationship (Hepler & Woolley, 1973):

$$\ln K_w(T) = -(34865/T) + 939.8563 + 0.22645T$$
$$- 161.94 \ln T \qquad (3.1)$$

Although no measurements have been performed at subzero temperatures, the very good fit of eqn (3.1) over a large temperature range should make the extrapolation of $\ln K_w$ reliable even to the undercooled liquid range. Substitution of $T = 238$ K ($-35\,°C$), a temperature not uncommon in the natural environment, predicts that $pK_w \simeq 17$, i.e. $a_{H^+} = a_{OH^-} = 3.2 \times 10^{-8}$ g ion dm^{-3}. The degree of ionization of water thus decreases quite markedly with a decrease in temperature.[†]

Since the ionization of an acid or a base in aqueous solution is expressed relative to the ionization of the solvent itself, it is to be expected that K_a and K_b will also exhibit a temperature dependence. Figure 3.1 is a van't Hoff representation of acid dissociation constants as a function of temperature, normalized to pK_a at 30 °C. The very complex temperature dependence of pK_a is at once apparent. It appears that for each acid there is a particular temperature at which the acid strength passes through a maximum. For the carboxylic acids this appears to be in the region of 20 °C. Assuming that the $\ln K_a(T^{-1})$ relationship can be described by a parabola, then extrapolation to, say, $-20\,°C$, indicates that for malonic acid the temperature effect is small, but for glycine and $H_2PO_4^-$ it is large. For H_2CO_3 it is very large indeed; probably a parabolic extrapolation is no longer warranted in that case.

Figure 3.1 graphically demonstrates that in the preparation of buffer mixtures, especially those to be used at low temperatures, it is essential to allow for the changes in K_w and K_a due to temperature changes. However, such temperature sensitivity is also likely to affect the stability and interactions of polyelectrolytes, such as proteins, nucleotides and carbohydrates in solution. While the shifts in pK values of individual acidic or basic residues may not be large for, say, a 10 °C decrease in temperature, the cumulative effect of many such residues, coupled with the fact that electrostatic interactions are of a long range nature, may well produce significant changes in the conformational stability of a polyelectrolyte. We shall return to this subject presently.

[†] An understanding of the precise significance of the term pH and the pitfalls inherent in its measurement at temperatures other than 25 °C is crucial in biochemical or physiological experiments where small changes in hydrogen ion activity are believed to produce marked effects. For a very lucid discussion of this topic, the reader is referred to an essay by Taylor (1981).

Another property which exerts a profound influence on interionic forces in solution is the dielectric permittivity (ϵ). It is often claimed that many of the peculiarities of aqueous solutions derive from the high dielectric permittivity of water – 80 at 20 °C. This is by no means the whole truth, because there are organic solvents, particularly the N-monosubstituted amides (e.g. N-methylformamide), which have dielectric permittivities in

Fig. 3.1. The effect of temperature on pK_a of some acids used in the preparation of pH buffers, normalized to 30 °C. Reproduced from Franks (1984).

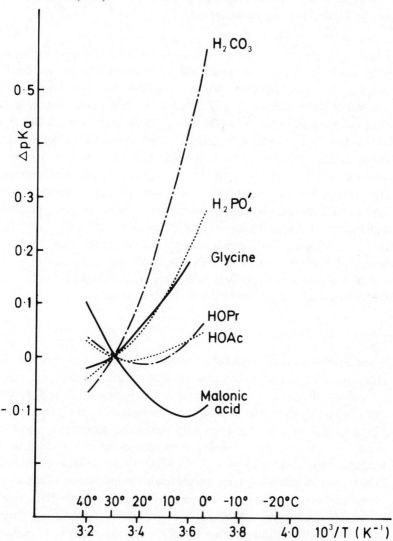

excess of 170 (Barthel, Gores, Schmeer & Wachter, 1984), but which do not exhibit the eccentric solvent properties of water. Usually a high ϵ arises from a large molecular dipole moment, but this is not the case with water which has a dipole moment of only 1.84 D. In comparison, acetone, with $\epsilon = 21.5$, has a dipole moment of 2.85 D. The high ϵ of water is the consequence of the particular association pattern in the liquid. The tetrahedral water structure becomes progressively better defined with decreasing temperature (if freezing is avoided), so that ϵ increases with decreasing temperature, reaching 100 at $-20\,°C$ (Hasted & Shahidi, 1976). The effect of ϵ on ionic equilibria in solution is expressed through the Debye–Hückel limiting equation

$$\ln \gamma_\pm = -Az_+z_-I^{\frac{1}{2}} \tag{3.2}$$

where $\gamma_\pm$ is the mean ionic activity coefficient, I is the ionic strength, z is the valence and A is a constant which contains the temperature, density and ϵ of the solvent medium. A change in the temperature from $+20$ to $-20\,°C$ is accompanied by a 6% reduction in $RT\ln\gamma_\pm$ which is the electrostratic part of the excess free energy of an ionic solution. This may not seem a major effect, but in many biological processes minor physicochemical changes are vastly amplified (see, for instance, the H/D substitution, discussed on p. 4). This is particularly true for a complex structure, such as a globular protein, which depends for its stability on the delicate balance between several different types of interactions, involving also interactions with the solvent (see Chapter 4).

Other homogeneous equilibria in aqueous solution which may be subject to complex temperature effects include solubility, especially in multicomponent solutions. Usually the solubility of solids and liquids increases with rising temperature according to the relationship

$$\frac{d\ln S}{dT} = \frac{\Delta H}{RT^2} \tag{3.3}$$

where S is the solubility and ΔH is the heat of solution. For most aqueous solutions ΔH is itself a function of temperature, so that the simple integration of eqn (3.3) is inappropriate. The solubility (or $\ln S$) is often expressed as a polynomial in T, in a similar manner to $\ln K_w$ in eqn (3.1). The solubility of gases increases with decreasing temperature, and the indications are that the solubility of oxygen increases threefold in the temperature interval $+20$ to $-20\,°C$. Many simple organic compounds exhibit complex solubility/temperature relationships, usually symptomatic of hydrophobic interactions, the origins of which have been briefly explained in Chapter 1. Minima in the solubility/temperature curves, known as lower critical solution temperatures, are peculiar to aqueous solutions of molecules with substantial apolar residues (Franks, 1975).

Behaviour resembling lower critical demixing is also observed with aggregation processes, such as micelle formation by surfactants (Kreshek, 1975), or peptide subunit aggregation to form microtubules, actin filaments or viruses (Lauffer, 1978). In every case there is a temperature below which the aggregation process is reversed and dissociation sets in. Lower critical demixing is also very common in aqueous solutions of polymers, such as polyethylene glycols, polyvinyl alcohol and polymethacrylamide (Franks, 1983a). The solid/liquid phase diagram of a series of polyethylene glycols is shown in Fig. 3.2; the lower critical temperature and solubility are seen to depend on the molecular weight (Burchard, 1983). In general, it would seem that undercooled water should be a very good solvent for all substances whose behaviour in solution is dominated by hydrophobic interactions.

Turning now to the dynamic properties of water and solutions at

Fig. 3.2. Solid/liquid phase diagram of polyethylene glycol/water mixtures with molecular masses ranging from 2.2×10^3 to 1020×10^3. Outside the loops the mixtures are homogeneous; inside the loops they exist as two phases in equilibrium. After Burchard (1983).

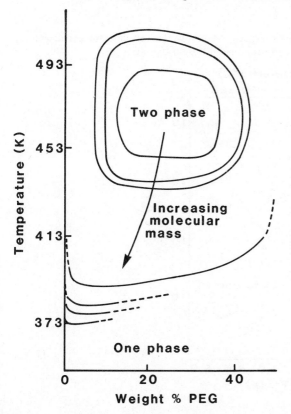

subzero temperatures, we find that bulk viscosity and diffusion are markedly affected. The various properties are related by the equation

$$t = \langle r^2 \rangle / 12D = kT/6\pi\eta a \tag{3.4}$$

where $\langle r^2 \rangle$ is the mean square displacement that a molecule of radius a undergoes in time t, D being the self-diffusion coefficient and η the shear viscosity. In fact, t is closely related to the characteristic times (t_r, t_D) in Fig. 2.7. In common with most other properties of water, both t and η show a progressively steeper increase with an increasing degree of undercooling. Figure 3.3 summarizes the various quantities in eqn (3.4) (Franks, 1981b). It is immediately apparent that the simple Arrhenius equation, which relates D, t and η to temperature through a temperature-independent energy of activation, is an unacceptable approximation in the subzero temperature range. The activation energy itself increases rapidly with decreasing temperature, as is apparent from the curvature of the plots. Since diffusive motions play an important part in physiological processes, the temperature dependence of the dynamic properties shown in Fig. 3.3 may well have a bearing on overwintering phenomena.

The effect of temperature on the kinetic coupling of reactions is another

Fig. 3.3. Dynamic properties of liquid water: self-diffusion coefficient (D), viscosity (η) and rotational diffusion time (t).

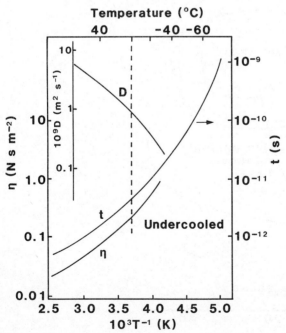

important aspect of physiological stress. We have already briefly touched upon the probable uncoupling produced on a series of reactions by the substitution of H_2O by D_2O. A similar effect could be produced by a change in the temperature where the individual steps in a reaction chain have different energies of activation. Consider the reaction scheme

$$A \to B (k_1)$$
$$B \to C (k_2)$$

where B can be regarded as an intermediate in the production of C. If A_0 is the initial concentration of A, then the concentration of C is given by

$$C(t) = A_0[1 + (k_1 - k_2)^{-1}(k_2 \exp(-k_1 t) - k_1 \exp(-k_2 t))] \quad (3.5)$$

The corresponding expression for the intermediate B is

$$B(t) = A_0 k_1 (k_2 - k_1)^{-1}(\exp(-k_1 t) - \exp(-k_2 t)) \quad (3.6)$$

Figure 3.4 shows the composition–time curves for the two reactions with $k_1 = 0.05$ min^{-1} and $k_2 = 0.1$ min^{-1}. There is only a short induction period in the build-up of product C, and the concentration of the intermediate B never rises above 25% and reaches an almost stationary state after approximately 30 minutes. In order to study the effects of temperature on

Fig. 3.4. Composition–time curves for simultaneous reactions $A \to B$ and $B \to C$, with the rate constants indicated, at 25 °C. $B(t)$ and $C(t)$ are given by eqns. (3.5) and (3.6).

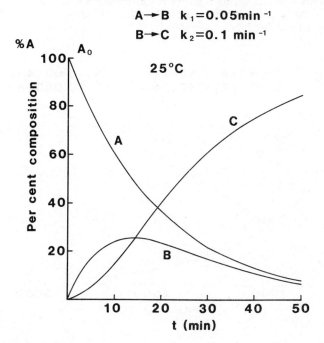

$A \to B \quad k_1 = 0.05 \text{min}^{-1}$

$B \to C \quad k_2 = 0.1 \text{ min}^{-1}$

the reaction scheme, it is useful to transform eqns (3.5) and (3.6) with the aid of the dimensionless parameters $z = k_1 t$ and $y = k_2/k_1$. If $b = B/A_0$ and $c = C/A_0$, then

$$b = (y-1)^{-1}(\exp(-z) - \exp(-yz)) \qquad (3.7)$$

and

$$c = 1 + (1-y)^{-1}(y\exp(-z) - \exp(-yz)) \qquad (3.8)$$

If we also put $a = A/A_0$, then by plotting $(1-a)$ against b or c, for different values of y, it is possible to show the whole range of time from 0 to ∞, as $(1-a)$ goes from zero to unity, because $a = e^{-z}$. The time dependence

Fig. 3.5. Reduced kinetics for the reaction scheme shown in Fig. 3.4, according to eqns. (3.7) and (3.8).

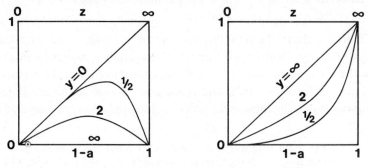

Fig. 3.6. Composition–time curves for the reaction scheme shown in Fig. 3.4, at $-20\,°C$, assuming energies of activation for the two reactions are 50 and 70 kJ mol^{-1} respectively.

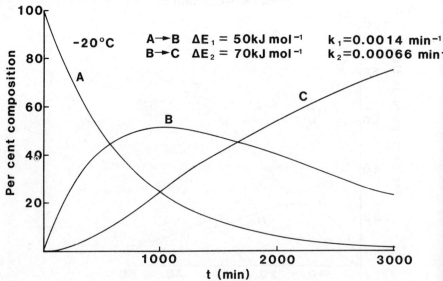

of b and c for different values of y is shown in Figs. 3.5a and b. The maximum value of B, as measured by b, depends on y and is given by

$$b_{max} = y^{y/(1-y)}$$

with b_{max} shifting to shorter times as y gets larger.

Let us now assume that $k_2/k_1 = y = 2$, as shown in Fig. 3.4, represents the optimum physiological state for a given pair of reactions at 25 °C. Let us further assume that reactions 1 and 2 have activation energies of 50 and 70 kJ mol^{-1} respectively. The time course of the two reactions at -20 °C is shown in Figs. 3.6 and 3.4; k_2/k_1 is now 0.47. Apart from the general slowing down in the disappearance of A, we note that there is a considerable enrichment in the intermediate species B at the expense of the product C, and that B only disappears very slowly. This simple example serves to demonstrate that for a system of coupled reactions, where the rate constants have their optimum values at the physiological temperature, but are subject to different activation energies, the kinetic coupling is easily perturbed. If the further complication is introduced where the intermediate species can simultaneously decompose by means of another pathway, then a change in the ratio of the various rate constants, produced by different activation energies, can produce chaos in the reaction cycle.

3.2   Freezing of dilute aqueous solutions

The previous section deals with the manner in which subambient temperatures can affect the properties of aqueous solutions in the absence of ice, that is, either at temperatures above T_f or in the undercooled state. When a solution is cooled to below its equilibrium freezing point and seeded with an ice crystal, ice will separate out as a pure phase and the solution concentration will increase, with a concomitant decrease in T_f. For a very dilute solution the simple relationship between freezing point and composition is

$$\Delta T_f = K_f m \tag{3.9}$$

where m is the solute concentration in mol (kg solvent)$^{-1}$ and K_f is the cryoscopic constant which depends solely on the physical properties of the solvent and is given by

$$K_f = R(T_f^0)^2/1000 \, \Delta H_f \tag{3.10}$$

ΔH_f being the latent heat of fusion. Equation (3.9) states that the freezing point depression of a solution containing 1 mol of solute in 1 kg of solvent is K_f °C, irrespective of the nature of the solute. Since the derivation of the equation is based on a number of simplifying assumptions, it can only be expected to hold for very dilute solutions and is therefore of limited practical use. A better approximation is provided by the use of mol fraction units of concentration. Where freezing point depressions exceeding 1 °C are involved, even this approximation is no longer adequate, so that more exact equations need to be employed. One alternative is to express the water activity a_w as a power series in ΔT_f and to allow for the fact that the latent heat is itself a function of temperature. A relationship which has the advantage of expressing $\ln a_w(\Delta T_f)$ in a closed form is

$$-R \ln a_w = \left(\frac{\Delta H_f + \bar{L}_1}{T_f T_f^0}\right) \Delta T_f - (\Delta C_f + \bar{J}_1)\left[\frac{\Delta T_f}{T_f} + \ln\left(1 - \frac{\Delta T_f}{T_f}\right)\right]$$

$$-\Delta\beta\left[\frac{\Delta T_f}{2T_f} - \frac{T_f^0 \Delta T_f}{T_f} + T_f \ln\left(1 - \frac{\Delta T_f}{T_f}\right)\right] \qquad (3.11)$$

In eqn (3.11) ΔC_f is the heat capacity of fusion, $\bar{L}_1$ and $\bar{J}_1$ the relative partial molal enthalpy and heat capacity of water, respectively, in the solution and $\Delta\beta$ is given by

$$\Delta C_f + \bar{J}_1 = \text{constant} + \Delta\beta T$$

to allow for the temperature and concentration dependence of ΔC_f and $\bar{J}_1$.

Since $\Delta\beta$ appears only in the third term, it requires freezing point determinations of the highest precision for this term to contribute significantly to $R \ln a_w$. The quantities $\bar{L}_1$ and $\bar{J}_1$ are obtained from measurements of the heat of dilution and its temperature coefficient. Reliable $\bar{L}_1$ data for aqueous solutions are now becoming available, but they are usually confined to 25 °C, and there are few direct measurements of $\bar{J}_1$, let alone $\Delta\beta$. Equation (3.11) is therefore of limited practical use. It is nevertheless important to realize that eqn (3.9) is inadequate for any solution outside the millimolar concentration range, but that its use for such very dilute solutions requires freezing point measurements of the very highest order of precision. Although it would be attractive to be able to calculate the course of liquidus (freezing point/composition) curves by equations such as (3.11), this has never yet been achieved, mainly through lack of the necessary thermodynamic data for concentrated solutions. In practice, freezing point determinations have been limited to molecular weight estimations and to the study of solute interactions in dilute solutions via the estimation of virial coefficients; see eqn (1.2).

3.3   Freezing of concentrated aqueous solutions

In principle, the mutual freezing point depressions in a two component system could also be calculated by the use of equations of the type (3.11), with the two curves meeting at the eutectic point, the temperature below which no liquid exists. As mentioned above, however, the necessary physical data are lacking, especially for the non-aqueous component. The plotting of phase diagrams is therefore performed experimentally, by the determination of cooling curves at different compositions. Although this is usually regarded as a simple procedure, the events that occur during the cooling of a heterogeneous mixture are quite complex (Franks, 1982c), so that a brief analysis of a typical curve serves a useful purpose.

A typical cooling curve for a simple binary mixture is shown in Fig. 3.7. The continuous line is the recorded temperature, with the bold portion corresponding to freezing, i.e. the accumulation of ice. Initially the liquid undercools, the degree of undercooling being governed by the probability of nucleation, as discussed in Chapter 2. Under normal circumstances nucleation will be caused by stray particulate matter, so that for samples exceeding a few microlitres, undercooling is usually limited to a few degrees. However, if adequate precautions are taken to prevent nucleation by stray impurities, e.g. by emulsifying the aqueous solution in an inert oil in the form of μm dimension droplets, considerable undercooling can be achieved, depending also on the concentration of the solution. Undercooling as a survival mechanism will be described in Chapter 6.

In Fig. 3.7, undercooling is terminated at A which is the nucleation

Fig. 3.7. Typical cooling curve; A is the nucleation temperature, AB corresponds to recalescence (the release of latent heat); C corresponds to the onset of non-equilibrium freezing; freezing is complete at D. The bold line corresponds to freezing. After Franks (1982c).

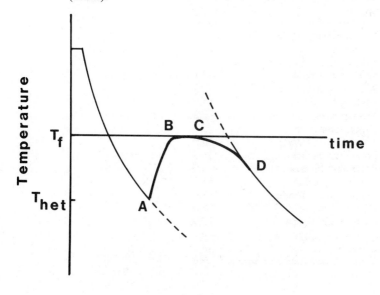

temperature. The release of latent heat causes the temperature to rise to the equilibrium freezing point T_f, where it is maintained for some time during which ice accumulates and the residual solution becomes freeze concentrated. This produces a local lowering of T_f which retards further freezing. Eventually the depletion of water, coupled with the competition for space, causes a retardation in freezing, so that the temperature will once again begin to fall at C. When the temperature reached D, freezing is complete and the cooling rate will then become dependent on the thermal conductivity of ice. At very high cooling rates the latent heat evolved is removed before the temperature can rise to T_f. Crystallization of ice is then partially inhibited.

The solid–liquid phase diagram is constructed from cooling curve data by plotting T_f as a function of concentration. Where both water and the solute can crystallize from solution and where no compounds (hydrates) are formed, the phase diagram takes the form of a simple eutectic system. Examples of such systems are given in Table 3.1. Below the eutectic temperature T_e the system consists of a mixture of two crystal types, the sizes of which are governed by the nucleation characteristics of the two components. An example of eutectic separation is shown in Fig. 3.8 for the system $H_2O + KCl$ (MacKenzie, 1977).

Table 3.1. *Eutectic temperatures of various solutes*

Solute	T_e (°C)
Monosodium citrate	−2.0
Disodium citrate	> −12.0
Trisodium citrate	−6.9
Sodium carbonate	−2.1
Sodium bicarbonate	−2.3
Sodium nitrate	−18.5
Monopotassium citrate	−2.2
Dipotassium citrate	−15.6
Tripotassium citrate	< −40.0
Potassium carbonate	−36.5
Potassium bicarbonate	−6.0
Calcium chloride	−55.0
Magnesium chloride	−33.6
Hydrochloric acid	−86.0
Sucrose, anhydrous	−13.95
D-Fructose dihydrate	−9.7
Glycerol	−46.5

Fig. 3.8. Electron micrograph of a 20% KCl solution after rapid quenching to -150 °C and freeze drying at -70 °C ($T_e = -11.1$ °C). Note the spherulitic crystallization of ice and the random crystallization (homogeneous nucleation?) of KCl. Reproduced from MacKenzie (1977).

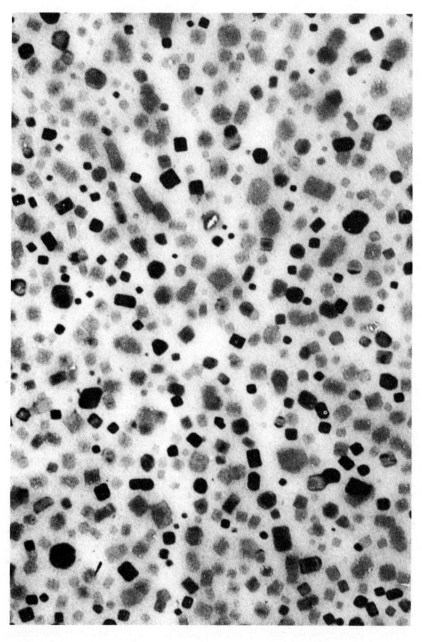

It is frequently observed that the solute does not readily crystallize from aqueous solution, in which event the phase diagram will not show a eutectic discontinuity. Instead, the liquidus curve will continue in a monotonic manner, characteristic of a supersaturated solution. It must be emphasized that such behaviour is indicative of thermodynamic meta-stability or instability.

The slope of the liquidus curve is a measure of the solute molecular interactions which are expressed by the higher order terms of eqn (3.11) through the quantities $\overline{L}_1$ and $\overline{J}_1$. Of course such interactions will also modify the hydration behaviour, but such modifications cannot be described explicitly from an analysis of the T_f vs. composition curve. In systems with pronounced solute–solute interactions the combined effects of low temperature and freeze concentration lead to a marked increase in the viscosity which retards both ice nucleation and crystal growth. As is shown in Fig. 2.7, the viscosity increases much more steeply than is predicted by the Arrhenius equation, and this steepness is further compounded when accompanied by an increasing solute concentration. The relationship between viscosity and diffusion, given by eqn (3.4), indicates that when η reaches 10^{14} N s m^{-2}, the time scale for diffusion is of the order of 10^6 s nm^{-1} (300 000 years cm^{-1}). Under such conditions the solution takes on the mechanical properties of a plastic solid and is referred to as a glass. It must however be emphasized that the system is not a crystalline solid but an undercooled liquid. The physical properties of metastable concentrated aqueous solutions at subzero temperatures are fundamental to the understanding of freeze tolerance and freeze resistance phenomena, and they will therefore be discussed in some detail.

3.4   Supersaturated solutions: metastable water

Because the freezing point of water – the essential chemical substrate for all forms of life – lies almost exactly at the centre of the temperature range which we associate with life, the properties of complex aqueous systems at subzero temperatures are found to be of great significance to acclimatization and resistance phenomena. Since ice is the stable phase of water in this temperature range, a nonfrozen system, whether liquid or solid, is metastable: undercooled and supersaturated; its lifetime depends solely on the probability of nucleation which, in turn, is a function of the diffusion rates of the various components. Figure 3.9 illustrates the various macroscopic states that can exist at subzero temperatures. The glass transition temperature (T_g) of water in the form of amorphous ice is believed to be in the neighbourhood of 120 K (-153 °C) but the homogeneous nucleation temperature (T_h) of undercooled liquid water is 233 K, so that in practice the glass cannot readily be prepared from liquid water by cooling. The conventional method used for the preparation of amorphous ice involves the deposition of water vapour onto a solid surface at < 100 K (Narten, Venkatesh & Rice, 1976). Claims have recently been made (Brüggeller & Mayer, 1980; Mayer & Brüggeller, 1983; Dubochet, Adrian & Vogel, 1983) for the successful vitrification of liquid water by ultrarapid cooling. This subject will be further explored in Chapter 9 because of its fundamental importance in cryofixation technology and low temperature electron microscopy.

Even if pure water cannot be vitrified by direct cooling to its T_g, the vitrification of aqueous solutions has been achieved (Franks *et al.*, 1977). The relationship between temperature, cooling rate and concentration is complex, and the reader is referred to specialist publications (Diller & Lynch, 1983, 1984*a*, *b*). There are, however, several general rules: the nucleation temperature of water is depressed by solutes, with $\Delta T_h \simeq 2\Delta T_f$ at any given concentration (Rasmussen & MacKenzie, 1972; Franks, 1981*c*). The glass transition curve of a binary mixture is probably a monotonic function of the concentration, and T_g of a pure substance is a function of its molecular weight. The nucleation rate increases with decreasing temperature (see eqn (2.6)), whereas the rate of crystal growth increases with increasing temperature. The latent heat (i.e. the driving force) of crystallization of water decreases in a nonlinear manner with decreasing temperature. The values at T_f and T_h are 6 and 3.75 kJ mol^{-1}, respectively (Franks, 1982*c*). When the latent heat reaches zero, the liquid is said to be hypercooled.

The degree of undercooling/freezing at subzero temperatures can be controlled through the following variables: temperature, concentration and rate of cooling. Under natural environmental conditions only the concentration of protectant solutes in the tissue fluids can be controlled by the organism. An analysis of a typical temperature/concentration state diagram may help to explain the various metastable states that can exist.

Fig. 3.9. Forms of stable and metastable water which can exist at subzero temperatures.

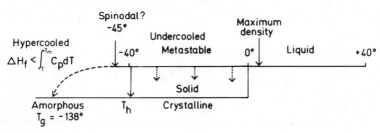

Figure 3.10 is drawn from the available experimental data for the system water–sucrose (MacKenzie, 1977). The bold lines are the freezing point (liquidus) and solubility (T_s) curves which meet at the eutectic point, T_e. These two curves are the only two equilibrium coexistence lines in the diagram. The T_s curve terminates at the melting point of pure sucrose, 465 K. Sucrose does not readily crystallize at T_e, so that in practice the T_f curve continues past T_e but now is indicative of a supersaturated solution. The glass transition curve connects T_g of water and sucrose (325 K) (Soesanto & Williams, 1981). As explained above, it is the locus of the temperature/composition curve at which the viscosity is $\sim 10^{14}$ N s m^{-2}, and below which the system can be regarded as an amorphous solid, since crystallization within a reasonable period of time becomes impossible.

The T_h curve is the lower limit of undercooling in the absence of stray impurities and at reasonable cooling rates. Thus, within the area bounded by T_h and T_g the solution will be nucleated and vulnerable to freezing. The growth of crystals may be prevented by rapid quenching, in which case freezing is likely to occur during rewarming, usually at T_d, the devitrification temperature. Unless precautions are taken to achieve maximum under-

Fig. 3.10. Water–sucrose composition/temperature state diagram. Bold lines represent equilibrium phase coexistence curves. Other lines denote processes under kinetic control. Broken lines indicate probable (undetermined) relationships.

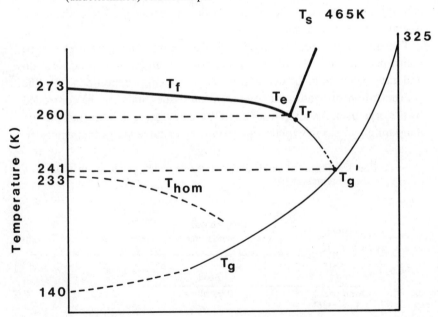

cooling, the solution will freeze at some temperature between T_f and T_h, catalysed by stray particulate matter. When this occurs, the evolution of the latent heat causes the temperature to rise to T_f and, provided that the cooling rate is moderate, the composition of the mixture will then follow the T_f curve, right into the region of supersaturation. Where the T_f and T_g curves meet, at T'_g, ice crystallization will cease. The system will then be a mixture of the solid supersaturated solution with ice embedded in it. The ice crystal size distribution depends on the thermal history of the mixture. Substantial undercooling gives rise to a highly nucleated state, and thus a high concentration of small crystals. Since the surface: volume ratio in such a system is large, it corresponds to an unstable state. Small crystals have a higher sublimation pressure than large crystals, the excess vapour pressure being given by the Kelvin equation

$$RT \ln (p/p_0) = 2M\sigma/\rho r$$

where p is the sublimation pressure of a (spherical) crystal of radius r, p_0 is the equilibrium sublimation pressure, M is the molecular weight, ρ is the density and σ the ice/solution interfacial tension. The rate at which small crystals can sublime and larger ones grow depends on the viscosity of the continuous phase. Above T_d the process, known as recrystallization, begins, but it becomes rapid only at some higher temperature, referred to as the recrystallization temperature, T_r, indicated in Fig. 3.10. T_r is of course independent of the initial concentration of the solution, because once the solution has frozen (either during cooling or warming), its concentration will correspond to a point on the T_f curve, i.e. it will have been freeze concentrated. The importance of T_r is that the changes in crystal dimensions and shapes markedly affect the mechanical properties of the system, usually to its detriment. Recrystallization is therefore an important factor in long term, low temperature storage.

To regard T_g as a property of the binary system is probably something of a simplification. It is taken to be a measure of the temperature at which diffusion rates become measurable, see eqn (3.4). However, the glassy state contains two kinds of molecules; in Fig. 3.10 they are water and sucrose. They have very different molecular weights and require different amounts of energy to perform rotational diffusion. It is therefore not inconceivable that when a binary glass is warmed, water molecules will become mobile before the larger sucrose molecules can begin to rotate. There are, indeed, indications that two temperatures can thus be identified, the lower one corresponding to T_g in Fig. 3.10 and the higher one in the neighbourhood of 233 K, possibly associated with the onset of diffusional freedom of the sucrose molecules (MacKenzie, 1977; Reid, 1979).

The T_g curve provides a rather artificial termination to ice crystallization at T_g'. The water content of the mixture corresponding to T_g' is frequently referred to as the unfreezable water and is of great biological and technological significance. In the water–sucrose system it amounts to 0.56 g water/g sucrose. It must be emphasized that this particular ratio is fortuitous, because there is no theoretical relationship between $T_f(c)$ and $T_g(c)$, c being the concentration. The liquidus curve is determined, through eqn (3.11), by the potential of mean force between hydrated sucrose molecules, whereas T_g is determined by the diffusional properties of the molecules in the system. Nevertheless, the unfreezable water in a system is frequently said to be 'bound' water (Biswas, Kumsah, Pass & Phillips, 1975), in the sense that a given number of water molecules are thought to be bound to a solute molecule and are for that reason prevented from freezing. There is even talk of 'strongly bound' and 'weakly bound' water molecules, although the possible origin of strong binding of water molecules is never explained (Ling, 1979). It cannot be emphasized too strongly that stoichiometric binding of water molecules by polar groups on organic molecules is most unlikely and there is no experimental evidence in support of such claims. It is of course possible to express the proportion of unfrozen water in terms of a so-called binding constant, but such a procedure is not based on reality and cannot stand up to a rigorous thermodynamic analysis..

Hydration is much better defined in ionic solutions, where the nature of the primary ionic hydration shell has been clearly characterized by neutron diffraction techniques (Enderby & Nielson, 1979). For example, in solutions of NaCl both the Na^+ and the Cl^- ions are surrounded octahedrally by six water molecules which could be regarded as 'bound', if it were not known that the residence time of such water molecules at the hydration site is of the order of 0.1 ns. Nevertheless, some biological text-books refer to ionic hydration as water binding. The futility of such an approach is demonstrated by the fact that when a NaCl solution is cooled, ice separates out, until at 252 K a simple eutectic is observed and the salt hydrate $NaCl.2H_2O$ then begins to crystallize, followed eventually by the anhydrous salt. Despite the fact that ion–water interactions are stronger than organic —OH–water interactions, there is no evidence for bound or even unfreezable water in the water–NaCl system.

Unfreezable water is the manifestation of a purely kinetic phenomenon which must not be confused with an equilibrium binding process. It is symptomatic of metastable, supersaturated systems in which the water has been robbed of its diffusional freedom and is thereby prevented from crystallizing. It must be remembered that at temperatures below T_f, ice is the stable phase, and the lower is the temperature, the larger is the *thermodynamic* driving force for ice crystallization. The fact that unfrozen water is able to coexist with ice is indicative of the large activation barrier preventing its diffusion to, and incorporation into, the ice crystal surface.

The phenomenon of unfreezable water is also related to the reluctance or inability of the *solute* to crystallize from aqueous solution, and we can now review the types of organic molecules that exhibit such behaviour. Of greatest importance, at least as far as *in vivo* cold resistance is concerned, are the polyhydroxy compounds, specifically carbohydrates and polyols (Franks, 1983*b*). Few sugars crystallize spontaneously, and among the common polyols only mannitol will crystallize readily from aqueous solution. For the remainder, supersaturated solutions are stable at quite low temperatures for long periods. Other groups of substances that are easy to undercool include globular proteins and some synthetic polymers, in particular polyvinyl derivatives and chemically modified starches and celluloses. The phenomenon of unfreezable water can also be observed in dispersions of supramolecular particles, e.g. cell suspensions or tissues. This is shown in Fig. 3.11 for suspensions of red blood cells, yeast cells and muscle tissue (MacKenzie, 1975). Although a colligative freezing point depression has no meaning here, it is seen that for all systems shown, the freezing point of the aqueous phase (distilled water in most cases) decreases steeply with increasing solids content at water contents of less than 50%. After freeze concentration to a solids content of 70–80% the freezing point tends to become invariant. This implies that between 20 and 30% of the water in the systems remains unfrozen, however low the temperature. Since one cannot speak of a glass transition in such systems, the water is probably rendered unfreezable by virtue of being confined in very small pores and capillaries in the solid substrate.

The state diagrams depicted in Figs. 3.10 and 3.11 are obtained from experimental measurements performed under conditions that allow the establishment of steady states, despite the fact that the solutions under test are supersaturated and not in stable equilibrium with ice. In other words, the liquidus curve describes the freezing of a system that has been seeded with ice and is cooled slowly enough for ice to crystallize under pseudo-equilibrium conditions. In the natural environment, where temperature changes resulting from climatic fluctuations are slow, freeze concentration probably follows the course shown by the $T_f(c)$ curves. On the other hand, when the cooling rate becomes an experimental variable, then it must be considered in relation to the rate of ice crystal growth. Because of its mainly technological importance, this topic will be discussed further in Chapter 9, and only the general factors that influence the rate of crystallization of ice from supersaturated solutions are considered at this stage.

The analysis of crystallization rates is conveniently performed in terms of the so-called TTT curve (time–temperature–transformation). This relates the time taken to crystallize a given fraction of the undercooled liquid to the temperature. Figure 3.12 shows typical TTT curves for the crystallization of ice from aqueous LiCl solutions (MacFarlane, Kadiyala & Angell, 1983). Experimentally the crystallization rates are measured by quenching the solution to some predetermined temperature T and determining the time taken for the ice to crystallize at that temperature, either by monitoring the latent heat of crystallization or by microscopic observation. The volume fraction $\phi(T)$ which crystallizes in time t is given by

$$\phi(T) = \tfrac{1}{3}\pi J u^3 t^4 \tag{3.12}$$

where J is the rate of nucleation and u the rate of crystal growth, both of which are functions of the viscosity (Avrami, 1941). Equation (3.12) indicates that the cooling rate is insensitive to ϕ, because $t(T)$ varies as $\phi^{\frac{1}{4}}$.

Fig. 3.11. Freezing temperatures of concentrated aqueous suspensions of (a) yeast cells, (b) glutaraldehyde fixed yeast cells, (c) bovine psoas muscle, (d) whole human blood and (e) glutaraldehyde fixed human erythrocytes. After MacKenzie (1975).

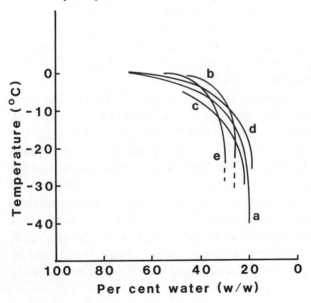

In Fig. 3.12 the upper branch of the curve describes nucleation limited crystallization, while in the lower branch the crystal growth rate determines the freezing process. The TTT curves also show the extreme sensitivity of crystal growth to the solution concentration (viscosity): thus, a concentration increase of only 1.5% slows down the crystallization by almost two orders of magnitude.

3.5   Equilibria and kinetics in part frozen systems

Because of the complexities associated with the crystallization of ice, very few aqueous solutions will freeze without the occlusion of pockets of supersaturation (constitutional undercooling). In order to achieve complete eutectic phase separation, it is necessary to take extreme precautions against undercooling, by seeding the solution with ice and using very low cooling rates. Even then, the solutes may be reluctant to crystallize at the eutectic temperature. The discussion in the previous section also suggests that the phenomenon of unfreezable water is very common, even in solutions of simple compounds; it is even more prevalent in highly heterogeneous systems. Closely related are the physio-chemical and biochemical changes which can occur while an aqueous system is subjected to freezing and during storage of such a frozen system.

Let us first trace the freeze concentration that takes place when a simple salt solution is cooled. Red blood cells are usually suspended in a phosphate buffered NaCl solution which is isotonic with the cell contents;

Fig. 3.12. Time–temperature–transformation (TTT) curves of ice crystallization in concentrated LiCl solutions, according to MacFarlane, Kadiyala & Angell (1983).

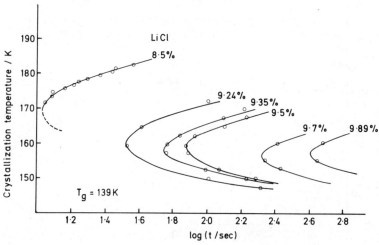

this isotonic concentration is 0.15 M, and the solution has an equilibrium freezing point of -0.54 °C. As the solution is cooled below this temperature, ice separates out and the residual liquid phase becomes more concentrated, until the eutectic temperature of -23.13 °C is reached; the salt concentration is then 4.7 M. The freeze concentration factor is therefore 313. The freeze concentration effect is illustrated in Fig. 3.13. It is seen that a tenfold concentration increase is already reached at -3 °C; at -10 °C well above 90% of the liquid solution has been converted to ice.

The concentration of a solution at a given subfreezing temperature is of course the same, whatever its initial concentration. It follows therefore that the freeze concentration effect is more pronounced for an initially dilute solution than for a more concentrated solution. This is best illustrated by means of an example (Fennema, 1975): consider one litre of a solution of concentration z mol l^{-1} which has been frozen to a concentration of $6z$ mol l^{-1} (a freeze concentration factor of 6). It now consists of $\frac{1}{6}$ litre of liquid phase and $\frac{5}{6}$ litre of ice. Starting with a solution which initially contains $3z$ mol l^{-1}, the concentration factor would be 2 (at the same temperature) and the system would then contain equal volumes of ice and liquid. In cases where freeze concentration is deleterious, it would be preferable to start with a concentrated solution. This should be an important consideration in processes like freeze drying.

Freeze concentration often affects the buffering ability of standard buffer mixtures, and this can have dramatic consequences. Commonly used phosphate buffer systems are subject to significant pH changes which are due primarily to freeze concentration; the influence of temperature alone can of course be calculated from a knowledge of the temperature dependence of K_w and K_a, as discussed in section 3.1. With mixtures of

Fig. 3.13. Freeze concentration of a 0.154 M NaCl solution (isotonic saline, used as a suspension medium for erythrocytes). For details see text.

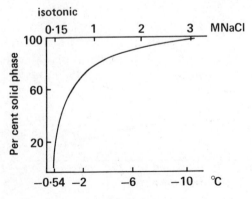

$H_2PO_4^-$ and HPO_4^{2-} it is found that the sodium salts suffer pH shifts of more than one unit as a result of freeze concentration; the potassium salts provide more stable buffer systems.

The observed pH shifts can be caused either by concentration effects as such or, in more extreme cases, by the actual crystallization of one or more of the buffer components under eutectic conditions. In phosphate buffers the crystallization of HPO_4^{2-} causes a decrease in the pH, whereas the crystallization of $H_2PO_4^-$ results in an increase. The sodium phosphate buffer mixture has a ternary eutectic point at -9.9 °C corresponding to a eutectic mol ratio $H_2PO_4^-:HPO_4^{2-}$ of 16.7:1. Any mixture that contains these salts in any other proportion must suffer pH shifts as one or the other salt crystallizes during cooling, until the ternary eutectic composition is reached. The corresponding mol ratio for the potassium double salt–ice eutectic is 0.48 at -16.7 °C. The lower eutectic temperature of the potassium phosphate buffer system accounts for the observation that under normal freezing conditions employed in biochemical studies or food processing, potassium buffers exhibit a better pH stability than the corresponding sodium phosphate buffer mixtures.

The presence of salt which might be added in order to control the ionic strength of a reaction medium will produce changes in the above behaviour, because there is now the possibility of quaternary eutectics with different mol ratios of the buffer components. It will be readily appreciated that pH shifts of the type described may give rise to secondary effects in systems which are stable only within limited pH ranges.

The most dramatic effects of freeze concentration are observed in the resulting kinetic changes. Freezing produces extreme deviations from normal Arrhenius behaviour, although the directions of such deviations cannot easily be predicted; they depend on the type of reaction considered. For reactions in homogeneous media freeze concentration is usually accompanied by rate enhancement, whereas with reactions involving heterogeneous steps, e.g. membrane-bound enzymes, the freeze induced discontinuity in the rate can take the form of either a retardation or an enhancement. Of the few reactions that have been subjected to a quantitative analysis, the mutarotation of glucose provides the best example (Kiovsky & Pincock, 1966). The eutectic temperature of the system ice–α-glucose monohydrate is -4.93 °C, and at -4 °C the residual liquid phase contains 27% glucose. For a second order reaction

$$A + B \rightarrow \text{products}$$

the rate of conversion of A is given by

$$d[A]/dt = -k_2 c_T([A_0][B_0])/c_0 \qquad (3.13)$$

where k_2 is the second order rate constant at temperature T, where $T > T_f$, c is the total number of mols of *all* solutes present, and the subscripts T and 0 refer to temperature T at which freeze concentration has taken place, i.e. $T < T_f$, and to the thawed state, respectively. In eqn (3.13) $k_2 c_T$ is the rate constant for the reaction when freeze concentration has taken place; both quantities depend on the temperature. Thus k_2 decreases with decreasing temperature according to the Arrhenius equation and c_T increases with decreasing temperature. For a dilute solution, where c_0 is small, freeze concentration is pronounced. Initially, therefore, $k_2 c_T$ increases with decreasing temperature. Eventually the freeze concentration effect becomes small compared to d ln k_2/dT, and the reaction rate then goes through a maximum and declines with decreasing temperatures. At high c_0 the observed rate enhancement due to freeze concentration is not pronounced. Equation (3.13) also shows that the reaction rate in the undercooled state differs from that in the part frozen state (at the same temperature) by a factor c_T/c_0. Therefore by changing this quantity, say by the addition of 'inert' solutes, it is possible to alter the degree of rate enhancement due to freeze concentration. It is assumed that low concentrations of 'inert' solutes do not affect the reaction rate in undercooled media but do affect freeze concentration. High concentrations of solute also affect the rate in the homogeneous undercooled medium through their influence on the water activity a_w.

Fig. 3.14. The effects of freeze concentration on the kinetics of mutarotation of glucose, after Kiovsky & Pincock (1966). The theoretical curves, given by eqn. (3.14), are shown as broken lines.

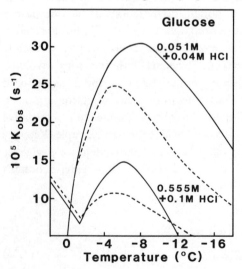

One further complication that may arise during freeze concentration is a change in the order of a reaction. In practice, many reactions in aqueous media are found to be pseudo-first order because the solvent, water, is present in large excess, i.e. $[A_0] \gg [B_0]$. After substantial freeze concentration, however, the concentration ratio changes by orders of magnitude and the reaction then has all the features of a true second order reaction. This can be shown by reference to the acid catalysed mutarotation of glucose under conditions of partial freezing. The kinetics are expressed by

$$k_{\text{observed}} = k_1 + k_2 \, c_T [\text{H}^+]_0 / c_0 \tag{3.14}$$

where k_1 is the rate constant for the spontaneous (uncatalysed) reaction which is not greatly affected by freezing. However, the order with respect to H^+ does change, due to compensating changes in the reaction volume and concentration. Figure 3.14 illustrates the effects of temperature and freeze concentration on the reaction for two c_0 values; the theoretical curves, as calculated from eqn (3.14) are also included. The effects of c_0 on rate enhancement are clearly indicated. In practice they can be even more dramatic: freeze concentration and constitutional undercooling can result in supersaturated solutions, in which the reactants are concentrated in 0.1% of the original liquid volume.

The complexities caused by freeze concentration can be further compounded in multicomponent systems with binary and ternary eutectics or systems containing macromolecules subject to temperature and concentration denaturation. Furthermore, in highly viscous systems diffusion may become the rate determining factor in chemical reactions and crystallization processes. A rigorous quantitative analysis of rate processes in such complex systems is impossible, but computer model building methods have been quite successful in predicting the behaviour of such systems during freezing.

4

Cryobiochemistry – responses of proteins to suboptimal temperatures

4.1   Scope and definitions

The previous chapter dealt with the influence of low temperatures on so-called simple chemical systems; it was seen that low temperature can exert a profound influence on equilibria and kinetics. We shall now examine the impact of these changes on the rather more complex biochemical processes. The discussions will be confined to *in vitro* processes at this stage, to be further developed and applied to intact cells and living organisms in the following chapters.

One of the basic tenets of molecular biology is the relationship between macromolecular structure and biological function. Thus, correct physiological functioning depends critically on the detailed structures and conformations of the molcules of life: proteins, nucleotides, lipids and carbohydrates, and on the interactions between the various species. This structure/function correlation has become firmly established in the teaching of life science subjects.

Most biological structures are believed to depend for their stability on noncovalent interactions of various types: electrostatic, dipolar, dispersion, hydration and hydrophobic (Kauzmann, 1959; Franks & Eagland, 1975; Pain, 1979; Finney, 1982). Although in a macromolecular system such contributions might individually be associated with quite large energies, biologically active structures are nevertheless very sensitive to temperature changes. This structure/temperature relationship will presently be explored in more detail.

Correct biological functioning is also sensitively attuned to the kinetics of coupled reaction sequences, such as the series of reactions involved in the oxidation of carbohydrates to CO_2 and H_2O (glycolysis). The analysis of the temperature dependence of two consecutive reactions discussed in the previous chapter (section 3.1) provides an indication of the complexities that can arise. It is probable, therefore, that the differential retardation of the various steps in a series of enzyme catalysed reactions, resulting from exposure to a suboptimal temperature, will have a profound impact on physiological function.

By the same token temperature can be used as a tool to probe the mechanistic details of sequences of coupled reactions. This has been done to good effect by Douzou and his colleagues. The term cryoenzymology, coined by him, now stands for the study of enzyme catalysed reactions at subzero temperatures in mixed aqueous/organic solvents to preclude freezing (Douzou, 1977). These techniques have been remarkably successful in allowing intermediate reaction species to be identified and isolated. In principle the study of a complex system at low temperature is more likely to produce useful information than could be obtained from high temperature experiments, because of the identification of temperature (kinetic energy) with uncertainty and disorder. Although cryoenzymology in exotic solvents, as currently practised, may be only of limited relevance to *in vivo* function, a short account is included in this chapter because of its innovative nature.

4.2 The thermal stability of native proteins

Proteins are macromolecules which are synthesized on the ribosome by the addition polymerization of amino acids. From a chemical point of view they are fairly simple polymers because the links between all the monomer residues are identical peptide bonds: $-CO.NH-$. Their complexities originate from the variety and versatility of the constituent amino acids, but all this variety resides in the amino acid side chain R of the molecule $NH_2CHR.COOH$. Amino acids can be classified according to the nature of the R group (North, 1979). Some acids have terminal ionogenic groups $-NH_2$ (e.g. lysine) or $-COOH$ (e.g. glutamic acid), others are classed as polar (serine, threonine) and yet others are classed as apolar (valine, phenylalanine). Proline and hydroxyproline take up a special position because, lacking the $-NH_2$ group, they are unable to form the normal amide type bond; instead they form an imide bond $-CO.N<$. A strict division along the above lines is impossible, because several amino acid side chains have both polar and apolar characteristics (e.g. tryptophan).

In marked contrast to the chemical similarity of proteins as polymers is their biological versatility. They fulfil a wide range of different functions, among them those of catalysts (enzymes), hormones, toxins, defence (immunoglobulins), transport, energy transduction (contractile systems) and structural frameworks (connective tissue). With the exception of the structural proteins, neither the amino acid composition nor the amino acid sequence provides obvious clues to the particular function of a given protein. Claims to the contrary are usually based on hindsight.

The biological activity of a protein is closely linked to its secondary and higher structures. Although these are determined by the primary amino acid sequence, the causal relationships between the peptide sequence and the details of folding and aggregation are not at all clear. What is clear is that the folded, *native* (active) state of a protein has a very marginal stability over that of the unfolded, *denatured* (inactive) state. For the commonly studied smaller globular proteins this free energy margin does not exceed 100 kJ mol^{-1} which is equivalent to the strength of about three hydrogen bonds (Finney, 1982). From X-Ray and neutron diffraction studies of protein crystals it is known that in the native state a medium sized protein contains upwards of 200 hydrogen bonds. If it is accepted that intrapeptide hydrogen bonds contribute to the stability of the native state, then it follows that some destabilizing factors must be at work so that the native state consists of a delicate balance between various opposing types of interactions. The balance is easily upset when one or several of these opposing interactions are strengthened or weakened. Such changes are produced by altering the physical and/or chemical environment of the protein (Franks & Eagland, 1975). Although in the laboratory environmental factors such as pressure, temperature, pH, ionic strength, or the chemical nature of the solvent medium can be altered at will, in the natural environment of most living organisms, temperature is the factor which most influences the stability of proteins; other *in vivo* effects are of a secondary nature, often triggered by changes in temperature. Although this discussion is concerned mainly with the effects of low temperatures, it must also touch upon the other factors that might contribute to the stability of proteins. Finney (1982) has attempted to draw up a 'balance sheet' of the various hypothetical contributions to the free energy of a folded protein relative to that of the same protein in some unspecified unfolded, denatured state. This free energy difference, $\Delta G(D \rightarrow N)$, which must appear as the bottom line of such a balance sheet, is the only quantity that has been measured with any degree of confidence. The contributing energies are either inferred from experiments with small molecule model compounds (Bigelow, 1967; Tanford, 1970; Shrake & Rupley, 1973) or calculated (Levitt, M., 1980). Yet other contributions, such as hydrophobic effects, which are expected to be highly significant, cannot even be calculated because of lack of a credible potential of mean force which could be used in such calculations.

Table 4.1 summarizes the essential features of Finney's balance sheet which refers to a hypothetical protein of the size of lysozyme or ribonuclease (approximately 120 residues) at *c.* 20 °C. It is seen that most of the contributing interactions are very much larger than $\Delta G(D \rightarrow N)$ but almost cancel to yield the marginal net stability. In particular, the contribution from hydrophobic interactions is quite uncertain, the magnitude depending on the particular model used in the calculation of the free energy gain associated with the transfer of an apolar residue from the aqueous solvent medium to the interior of the protein (Chothia, 1974; Franks, 1975; Richards & Karplus, 1980). By contrast, the stability derived from an intrapeptide hydrogen bond is quite small. Although not too much

Table 4.1. *Speculative balance sheet of factors contributing to the stability of a native globular protein containing 100 residues*

	Origin	Estimated magnitude (kJ mol^{-1})
Stabilizing factors		
ΔG (titration)	pK changes of acidic/basic residues	-20
ΔG (electrostatic)	Charge interactions at the periphery of the protein (minimized at isoelectric pH)	?
ΔG (hydrophobic)	Desolvation of apolar residues during folding, very sensitive to the *range* of the hydrophobic interaction	-150 to -2500
ΔH (hydrogen bonding)	Formation of intrapeptide hydrogen bonds	-300
$T\Delta S$ (water relaxation)	Release of water molecules (dehydration of peptide)	-500
Destabilizing factors		
ΔH (hydrogen bonding)	Distorted (non-linear) hydrogen bonds	$+400$ to 800
	Non-hydrogen bonded polar groups	$+250$
$T\Delta S$ (configurational)	Loss of configurational freedom (based on 8–20 J K^{-1} mol^{-1} per residue)	$+240$ to 800
Uncertain		
ΔH (van der Waals)	Changes in packing density; polar/apolar interactions	? ?
$T\Delta S$ (vibrational)	Changes in bond vibrational states (could be very large)	?
Net sum		
ΔG (D $\rightarrow$ N)	Experimental	-40 to -80

significance should be attached to the actual numbers, the principle of marginal stability is well established.

Matters become even more uncertain when we enquire into the probable influence of temperature on $\Delta G(D \rightarrow N)$. It seems to be axiomatic that an *ordered* structure, such as a folded protein, should become destabilized by an increase in temperature, and this is indeed observed. For any protein there exists a temperature T^* at which $\Delta G(D \rightarrow N)$ becomes equal to zero. In terms of the two-state model which prescribes a simple equilibrium between the two states, the equilibrium constant $K = [N]/[D]$ is equal to unity at T^*. For $T > T^*$, the protein will be predominantly in the denatured, inactive state.

The temperature dependence of ΔG yields the enthalpy, ΔH, and the entropy, ΔS, accompanying the denaturation according to the standard relationships

$$\Delta S = -d(\Delta G)/dT \quad \text{and} \quad \Delta H = [d(\Delta G/T)/d(1/T)]$$

However, both ΔS and ΔH may themselves be functions of temperature, according to

$$d(\Delta S)/dT = \Delta C_p/T \quad \text{and} \quad d(\Delta H)/dT = \Delta C_p$$

where ΔC_p is the change in heat capacity accompanying the denaturation.

From the above relationships it is possible to write the temperature dependence of ΔG as

$$\Delta G(T) = \Delta H[(T^* - T)/T] - \int \Delta C_p \, dT + T \int (\Delta C_p/T) \, dT \qquad (4.1)$$

where the first term on the right hand side refers to the transition, so that ΔH is the heat of denaturation at the temperature T^*, akin to a latent heat (Pfeil & Privalov, 1976). The detailed nature of the $\Delta G(T)$ curve is now seen to depend on the magnitude of ΔC_p and its possible temperature dependence. For $\Delta C_p = 0$, ΔG is linear in T. This corresponds to a native state which is stable for $T < T^*$, the margin of stability increasing with decreasing temperature. For $\Delta C_p \neq 0$, $\Delta G(T)$ is curvilinear, the curvature depending on the magnitude, sign and temperature dependence of ΔC_p. The situation is further complicated by the fact that temperature is not the only property that determines the magnitude of ΔG; others include dielectric permittivity, ionic strength and hydrogen ion activity (pH). Of these, only the ionic strength is independent of temperature. On the other hand, as discussed in Chapter 3, both dielectric permittivity and hydrogen ion activity are functions of temperature. It is not possible to solve for $\Delta G(T)$ analytically, and there are few experimental heat capacity studies on record of sufficiently high quality for $\Delta G(T)$ to be calculated with any degree of precision (Pfeil & Privalov, 1976). For chymotrypsinogen in the presence of HCl and neutral chloride the following empirical relationship fits the experimental data (Brandts, 1964):

$$[\Delta G(T)/4.18]\,\text{kJ mol}^{-1} = 121\,700 - 2226T + 11.57T^2 - 0.01783T^3$$
$$+ RT\ln\left(\frac{1 + a_{H^+}/5.0 \times 10^{-2}}{1 + a_{H^+}/3.2 \times 10^{-5}}\right)^{3.18}$$
$$- \nu RT\ln[\text{Cl}^-]/(0.01) \tag{4.2}$$

where ν is the number of ions produced by a molecule of the added salt. However, even this complicated equation takes no account of the temperature dependence of a_{H^+}.

In the absence of very high quality ΔC_p data, $\Delta G(\text{N} \to \text{D})$ is commonly measured by some biochemical or spectroscopic technique and the temperature derivates are obtained with the aid of the van't Hoff equation. An example of this type of procedure is shown in Fig. 4.1 for the denaturation of phosphoglycerate kinase (PGK), obtained from the thermophilic bacterium *Thermus thermophilus* and from yeast, respectively (Nojima, Ikai, Oshima & Noda, 1977). For various practical reasons the enzyme was studied in various concentrations of guanidinium hydrochloride. $\Delta G(T)$ was expressed in terms of a simple polynomial:

$$\Delta G(T) = AT + BT^2 + CT^3 + D \tag{4.3}$$

from which ΔH was obtained by differentiation. A comparison of the two

Fig. 4.1. ΔG and ΔH of denaturation of phosphoglycerate kinase from *T. thermophilus* (-----) in 2.26 M GuHCl and yeast (——) in 0.54 M GuHCl. Data from Nojima, Ikai, Oshima & Noda (1977). Extrapolated ΔG in the absence of GuHCl: *T. thermophilus*, 50 kJ mol^{-1} and yeast, 22 kJ mol^{-1} at 25 °C.

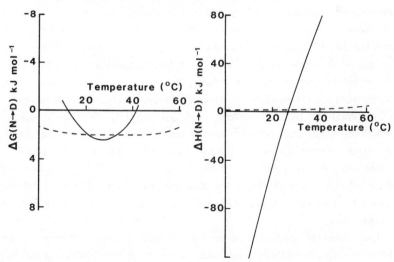

$\Delta G(T)$ curves shows up the qualitative similarity but quantitative difference between the same enzymes isolated from two different organisms. This difference is further magnified in ΔH. Whether the heat stability (and by inference, the cold stability) of the thermophilic PGK is due to the constant, almost zero enthalpy is open to speculation. The second temperature derivative of eqn (4.3) yields ΔC_p which is practically zero for *T. thermophilus* PGK but 7.3 kJ mol^{-1} for yeast PGK, causing the latter only to be stable in the limited temperature range 10–40 °C (in 2.26 M guanidinium hydrochloride).

The above results, based on fluorescence and circular dichroism measurements over a limited range of temperatures, are typical examples of biochemical investigations of protein denaturation. There must, however, be serious reservations about the use of the van't Hoff equation for the evaluation of ΔH and its higher temperature derivatives, based, as it is, on the simple two-state model of denaturation. Quite apart from that, the differentiation of experimental results which are themselves subject to significant variability can lead to very large uncertainties. Finally, the estimation of ΔH as well as ΔS from the same ΔG measurement provides for further errors (Krug, Hunter & Grieger, 1976). Apart from these criticisms which apply to most experimental work on record, there are further considerations that would apply to any experimental thermodynamic study. For instance, it is implicit in the use of eqns (4.1) or (4.2) that ΔG is independent of concentration, a condition that is not always adequately established by experiment. Another factor, not always carefully investigated, is the reversibility of the N→D transition. The whole thermodynamic edifice on which eqn (4.1) is based applies only to fully reversible processes. Under any other circumstances, the numbers obtained are meaningless. The nature of the unfolding process suggests that reversibility is most probably attainable only in very dilute solution, that is, under the very conditions where measurements are subject to large experimental errors. It is desirable, therefore, to choose experimental techniques that minimize such uncertainties. Preferably ΔH should be determined by an independent, calorimetric method under conditions which allow the extrapolation to zero concentration. The development of sensitive microcalorimeters, mainly by Privalov and his colleagues, has made this possible, as well as the direct measurement of ΔC_p. Heat capacity measurements over a range of temperature can then be used, in conjunction with eqn (4.1), to obtained ΔG at any given temperature (Pfeil & Privalov, 1979).

Unfortunately there are as yet no examples of such investigations on record for the low temperature behaviour of proteins, but the thermal (high temperature) denaturation of several proteins has been studied in great detail under conditions of true reversibility. For measurements of $C_p(T)$ the temperature scanning calorimeter is the most suitable tool; Fig. 4.2 shows the specific heat profiles of three lysozyme samples at different pH values (Privalov & Khechinashvili, 1974). Since constant scanning rates are employed, the abscissa is easily converted to temperature, while the ordinate is a measure of the specific heat.

The calorimeter output provides the following information:

(1) $C_p(T)$ of the native protein below the transition temperature T^*

(2) $C_p(T)$ of the denatured protein above the transition temperature T^*

(3) the heat capacity change ΔC_p which accompanies the N→D transition

(4) the enthalpy change ΔH^* corresponding to the transition

(5) the temperature of half conversion, T^*, where $K = 1$ $(\Delta G = 0)$ and

(6) the van't Hoff enthalpy change, according to the equation

$$\Delta H_{v.H.} = 2R^{\frac{1}{2}} T^{*\frac{1}{2}} \Delta C_p^{\frac{1}{2}} \tag{4.4}$$

Indirectly $\Delta G(T)$ is obtained from the data by the integration of eqn (4.1). The pH dependence of the transition further permits the estimation of the number of protons involved in the process.

Figure 4.3 shows the free energy surface as a function of temperature and pH of native and denatured lysozyme within the ranges 0–100 °C and pH 1–7 (Pfeil & Privalov, 1976). The curved nature of the iso-pH lines is

Fig. 4.2. Specific heat/temperature profiles of lysozyme as a function of pH. The area under the $C_p(T)$ curve is a measure of $\Delta H(\text{N}\to\text{D})$ and the change in the base is equal to $\Delta C_p(\text{N}\to\text{D})$. The width of the thermal transition is related to its cooperativity. After Privalov & Khechinashvili (1974).

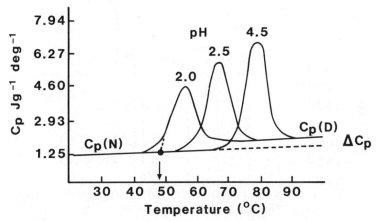

Fig. 4.3. *G*–pH–*T* surfaces of native and denatured lysozyme; drawn-out lines are contours of constant free energy. After Pfeil & Privalov (1976).

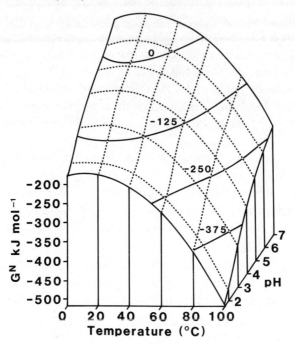

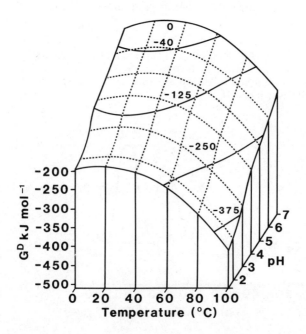

immediately apparent and is particularly pronounced for the D state. The pH isotherms are monotonic, but they exhibit inflexions near pH 4.

The free energy surface which describes the denaturation is obtained from the sum of the two surfaces in Fig. 4.3; it is shown for lysozyme in Fig. 4.4. The half-transition contour ($\Delta G = 0$) is drawn in bold. The complex nature of ΔG(pH, T) is at once apparent, as is also the curvature of $\Delta G(T)$ at constant pH. Lysozyme is a particularly stable protein; its low temperature behaviour cannot easily be inferred from the available free energy data which do not extend to subzero temperatures. Unfortunately lysozyme is the only protein for which separate $G^N(T)$ and $G^D(T)$ data are available, as are required for an analysis of the thermodynamic stabilities of the two states. For several proteins ΔG(pH, T) has been measured, and iso-pH sections through the surfaces of cytochrome c, chymotrypsinogen and the metmyoglobin–cyanide complex are shown in Fig. 4.5 (Pfeil & Privalov, 1979). Here again, the somewhat arbitrary cut-off at 0 °C makes it difficult to predict the full course of the ΔG contours.

However, the $\Delta G > 0$ contours in the cytochrome c surface do suggest the course of ΔG beyond the experimental range. The metmyoglobin–CN surface actually contains evidence of cold denaturation: at pH 12 there are two temperatures (0° and 50 °C) for which $\Delta G = 0$. The free energy profiles shown in Fig. 4.5 also indicate that for any given pH, $\Delta G(T)$ reaches a

Fig. 4.4 ΔG(N → D)–pH–T surface of lysozyme obtained from the algebraic sum of the individual G^N and G^D surfaces in Fig. 4.3. After Pfeil & Privalov (1976).

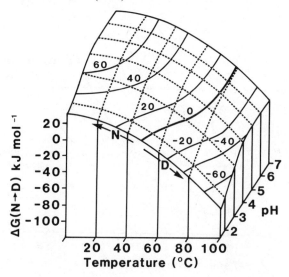

maximum value at some temperature. In the case of metmyoglobin–CN
at pH 12, $T_{max} = 25\ °C$ and $\Delta G_{max} = 13\ kJ\ mol^{-1}$.

While comprehensive, high quality thermodynamic data on protein
denaturation, of the type performed by Privalov and portrayed in Figs.
4.3–4.5, are very rare, there exists a body of data on thermal denaturation
referred to particular conditions of solvent composition, pH or ionic
strength. In most cases $\Delta G(T)$ takes the form of a parabola. One of the
most informative experimental studies is that by Brandts and his colleagues
of the denaturation of chymotrypsinogen; the stability profile is shown in
Fig. 4.6 (Brandts, 1964). T_{max} (12 °C) is clearly indicated and the data
between 0° and 60 °C are well fitted by the parabola, shown as the drawn
out line. The fitted equation can be differentiated to yield ΔH and ΔC_p.
The latter is surprisingly large: 12.6 kJ (mol K)$^{-1}$ at 40 °C, imparting a
large temperature dependence to ΔH and ΔS. For instance, ΔH changes
from 502 kJ mol^{-1} at 40 °C to zero at 10 °C and becomes negative at lower
temperatures. This behaviour can be contrasted with that of thermophilic
PGK, shown in Fig. 4.1. The large heat capacity effects which are
responsible for the free energy curves of the type shown in Fig. 4.6 appear
to be a common feature of many globular proteins. Unless the protein has
been artificially destabilized, e.g. by urea or guanidinium hydrochloride

Fig. 4.5. $\Delta G(N \to D)$/temperature profiles at constant pH; after
Privalov & Khechinashvili (1974). Broken lines denote the estimated
$\Delta G(T)$ behaviour in undercooled solution. Only in the case of
chymotrypsinogen has the cold denaturation in the undercooled state
been established experimentally (Franks, unpublished results.)

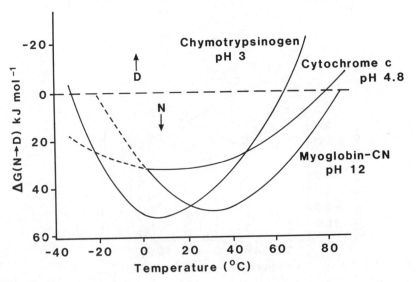

(see Fig. 4.1), the cold denaturation cannot usually be studied experimentally because freezing intervenes. Under special circumstances, however, where the nucleation of ice has been inhibited, the low temperature denaturation should become accessible to experimental study. By dispersing an aqueous solution of chymotrypsinogen in an inert carrier fluid (see chapter 2), we have been able to detect the reversible unfolding in the neighbourhood of $-33\ ^{\circ}C$ (Franks, unpublished results). As predicted by eqn (4.1), the $N \rightarrow D$ transition is exothermic ($\Delta H = -540\ kJ\ mol^{-1}$).

The phenomenon of cold inactivation is common in enzymes and other proteins which are composed of several peptide chains (Bock & Frieden, 1978). It often takes the form of a simple dissociation process in which the individual peptide subunits appear to maintain their native conformations but the molecular weight (sedimentation coefficient) of the intact aggregated structure is found to decrease during cooling. The cold induced dissociation may be partial, e.g. where a tetramer dissociates into dimers, or it may be complete. Paradoxically, cold inactivation of proteins can be the cause of low temperature injury, but the process can also be used by an organism as a means of protection against cold injury. These aspects will be further discussed in the following chapters.

Fig. 4.6. $\Delta G(N \rightarrow D)$/temperature profile of chymotrypsinogen at pH 1 and pH 3, according to Brandts (1964). The data can be adequately fitted by a parabola.

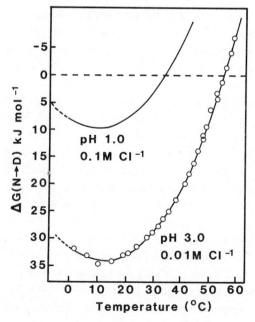

Well documented cases of low temperature enzyme inactivation include phosphofructokinase (Bock & Frieden, 1976), glucose-6-phosphate dehydrogenase (in erythrocytes) (Kirkman & Hendrickson, 1962), carbamyl phosphate synthetase (Guthohrlein & Knappe, 1968), ATPase (beef heart mitochondria) (Penefsky & Warner, 1965) and pyruvate carboxylase (Nakashima, Rudolph, Wakabayashi & Lardy, 1975). In all the above cases inactivation results from a dissociation which may also be accompanied by a minor conformational transition. Under the appropriate experimental conditions the inactivation is fully reversible. Some few enzymes, on the other hand, exhibit deactivation through aggregation at low temperatures, for example urease (Hofstee, 1949) and 17β-hydroxysteroid dehydrogenase (Jarabak, Seeds & Talalay, 1966), although in the latter case aggregation is preceded by a reversible subunit dissociation; the subsequent aggregation step is irreversible.

A similar sequence of reactions appears to take place with β-lactoglobulin which exists as an $\alpha\beta$-dimer at room temperature and pH 3–6. The dimer dissociates above pH 6 and this is followed by a conformational transition which, in turn, leads to nonspecific, irreversible aggregation. The various processes can be summarized as follows:

pH	3–6	6–9	9–12
Temp	20 °C	0 °C	
State	N	N$\rightleftharpoons$2R	2R$\rightarrow$2S $----\rightarrow$S$_n$

where N is the native state ($\alpha\beta$ dimer), R is the dissociated monomer, S is the denatured monomer and S$_n$ the aggregated state (Douzou, 1977). The initial dissociation can also be induced by cooling, with a transition pK of 7.7. The low temperature behaviour of the system can be further explored by replacing the aqueous solvent with 50% v/v ethane diol. This solvent does not appear to affect the conformational state of the protein subunits or the pK of the transition. The results are shown in Fig. 4.7. Here again the distinct curvature in p$K(1/T)$ should be noted, but also that the shift in pK due to cooling helps to protect the dimer in the pH range where it would dissociate at room temperature. The temperature mediated association/dissociation of large multimeric structures is of particular interest. Detailed studies have been carried out on tobacco mosaic virus coat protein (Lauffer, 1978) and on microtubules (Timasheff, 1978). The latter are composed of cylindrical aggregates of two types of globular peptide subunits (tubulin). The influence of temperature on the assembly reaction is shown in Fig. 4.8 and bears a striking resemblance to Fig. 4.7. Direct microcalorimetric studies have indicated a ΔC_p of -6.7 kJ (K mol)$^{-1}$, independent of temperature over the range investigated.

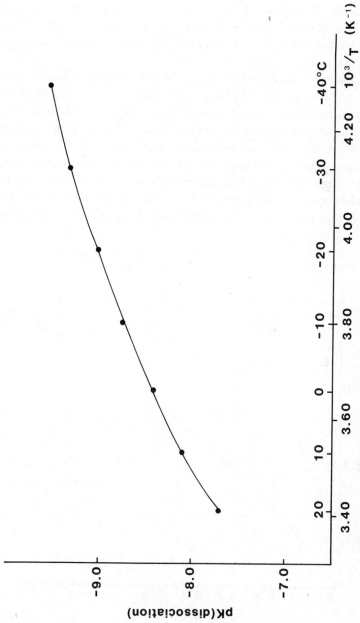

Fig. 4.7. Cold-induced dissociation of β-lactoglobulin in 50% aqueous ethane diol solution, according to Douzou (1977).

This is an excellent agreement with the van't Hoff value -6.3 kJ $(K\,mol)^{-1}$ derived from turbidimetric measurements, thus lending credence to the equilibrium model on which the estimation of the thermodynamic quantities is based. Here, as in the cases discussed earlier, the large ΔC_p is responsible for the pronounced curvature in $\Delta G(T)$ with indications of two dissociation temperatures. Although there are several other variables which affect the polymerization equilibrium, in the standard buffer mixture (phosphate buffer, pH 7), depolymerization of the tubules sets in on cooling to 20 °C.

Two general questions are raised by the denaturation/dissociation results: (1) what feature of the unfolding/denaturation/dissociation process is responsible for the large, positive ΔC_p values and (2) what might be the similarities and differences between the D states obtained by heat and cold induced denaturation. The answer to the first question is now fairly clear: the exposure to the aqueous solvent of previously buried apolar amino acid residues gives rise to large heat capacity effects. This has been verified with the aid of apolar model compounds and has its origin in the structural changes that take place in water in the proximity of apolar molecules or

Fig. 4.8. Reassembly of microtubules from tubulin subunits as a function of temperature. Redrawn from Timasheff (1978).

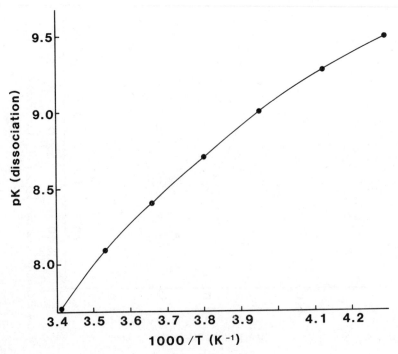

residues, referred to as hydrophobic hydration and described in Chapter 1. The two important features of hydrophobic hydration are that the enthalpy and entropy changes are extremely temperature sensitive and that over a certain temperature range the sign of ΔG is determined by the entropy (the last term on the right hand side of eqn (4.1)) rather than by the enthalpy, which would be a much more normal state of affairs. Even leaving aside these rather special effects which arise from the hydration of apolar groups in water, the D state would be expected to have a larger heat capacity than the N state. Heat capacity is a measure of the number of configurational degrees of freedom and this must be larger for a partly unfolded macromolecule than for its unique, native counterpart. However, the situation is not easily analysed because of the involvement of water in both the N and D states. The measured C_p values (and ΔC_p) include the contributions from hydration interactions and losses/gains in the configurational degrees of freedom of water in the presence of the protein.

The second question is much harder to resolve. The assumption of the simple two-state equilibrium demands that only one D state can exist, irrespective of whether it is approached by heating or cooling the N state. From a structural point of view this is tantamount to claiming that an ordered state can be disrupted by a reduction of the kinetic energy. Turning once again to the stability balance sheet in Table 4.1, we can speculate which of the contributions are likely to become weaker at low temperatures. The most obvious candidates are the hydrophobic interactions and the electrostatic repulsions (see eqn (3.2)). On the other hand, intrapeptide hydrogen bonds might be expected to become more stable at low temperatures. It is possible to speculate, therefore, that the main effect of low temperatures is the weakening of the tertiary structure, while the stability of the secondary structure is maintained. This hypothesis could also account for the cold induced depolymerization of many multisubunit structures. There is to date no experimental evidence to support such speculations. If true, they would require a reformulation of the denaturation process in terms of more than one D state, although the simple N → D model has served us well. In any case there is mounting experimental evidence that the denatured states obtained by heat, pH, urea and salt treatments are not identical. Indeed, the industrial processing of proteins to yield fabricated food products is based on this very fact. It is naive to equate the D state with that of a random polymer, i.e. one whose dimensions can be calculated by the laws of statistics.

Despite the experimental problems involved in subzero temperature studies of proteins, there is much to be gained from such investigations. In general terms, more useful information is likely to accrue from structural and dynamic studies where the kinetic energy in a system is reduced than from such studies where the level of the kinetic energy is increased; this would tend to disrupt the system in any case. Considering the functional properties of proteins, heat denaturation is largely irrelevant, except from an industrial processing point of view, since living organisms never have to survive temperatures at which their proteins would be heat denatured. On the other hand, temperatures associated with cold denaturation are of very real physiological significance, many overwintering organisms having to exist under, and acclimatize to, such conditions.

4.3   Enzyme reactivity, structure and kinetics at low temperatures

Although enzyme catalysed reactions obey the normal laws of kinetics, enzymes are rather special catalysts in view of their lability and chemical complexity. As discussed in the previous section, their structures in solution are sensitively dependent on temperature and the composition of the solvent medium. This, in turn, influences their effectiveness and specificity. Most major biochemical reaction pathways consist of a series of coupled reactions in which the product of any one reaction step serves as the substrate for the next one. Each individual step is chemically fairly simple, involving usually one of the four basic types of organic reactions: oxidation, reduction, hydrolysis and condensation.

An important principle of biochemical reaction pathways is the conservation of free energy: although the whole sequence may be accompanied by a large change in the free energy, each individual step is performed with a very small ΔG. This principle is well illustrated by the combustion of glucose to CO_2 and H_2O. In the many reaction steps involved in the conversion of glucose to pyruvate, the largest individual free energy change is -25 kJ mol^{-1} (the dephosphorylation of 2-phosphoenolpyruvate).

Under normal physiological conditions enzyme catalysed reactions take place fairly rapidly and the various intermediates exist at low stationary state concentrations. It is difficult then to analyse the mechanism of a reaction sequence in terms of the basic processes involved, e.g. proton transfer, ionization, breaking and making of covalent bonds, isomerization. To make possible such an analysis the slowing down of the reaction by a reduction in the temperature would seem attractive. In practice, subzero temperatures need to be employed for any benefit to be derived from such procedures, but this introduces all the problems associated with freezing. Douzou very elegantly overcame these problems by employing mixed aqueous/organic solvents with freezing points down to -100 °C. His methods of studying enzyme catalysed processes have become known as cryoenzymology (Douzou, 1977). The rationale is as follows: if reactions can be studied under homogeneous conditions (i.e. in a fluid medium) at subzero temperatures, then it should become possible to identify intermediate species and study the kinetics and thermodynamics of the separate reaction steps and hence eludicate the mechanisms of complex multistep processes. Given favourable circumstances, it might even be possible to isolate and purify labile intermediates, such as enzyme/substrate complexes, and to subject them to structural analysis by diffraction methods. Additional benefits of such low temperature diffraction procedures would be improvements in resolution and a reduction in radiation damage.

All the above aims have been realized, even to the extent that upwards of 30 crystalline proteins have been studied by X-ray diffraction at subzero temperatures in mixed solvent media (cryosolvents). An additional bonus obtained from diffraction measurements at a series of temperatures is information on the spatial fluctuations that occur within the macromolecule (Petsko, 1975). Such fluctuations about the time-averaged structure are likely to be highly significant for the biological function of a protein. In the case of myoglobin the mean square positional fluctuations of all the 1261 non-hydrogen atoms have been mapped. As might be expected, the atoms at the core, near to the haem group, are fairly immobile; dynamic freedom increases with distance from the centre. In particular, the atoms associated with reverse turns in the secondary structure possess a high degree of flexibility. It has thus become possible to map the degree of 'floppiness' of different regions in the macromolecule and to arrive at a 'dynamic structure' based, nevertheless, on diffraction measurements.

Even more spectacular successes have been achieved in the elucidation of reaction mechanisms and the characterization of transition states. The use of cryosolvents has become so popular among biochemists that the assumptions underlying the cryoenzymological procedures will be subjected to a brief analysis.

At the most basic level the following conditions must be met: it must be demonstrated that (1) neither the low temperature nor the cryosolvent adversely affect the structure and catalytic properties of the enzyme, and (2) there is kinetic correspondence between the reaction carried out at low and ambient temperatures. The latter condition implies a linear (or at least monotonic) Arrhenius plot. Actually the two conditions are closely linked, because kinetic correspondence can only occur if neither temperature nor solvent change give rise to major structural changes in the enzyme. The previously discussed stability/temperature relationships of proteins demonstrate that low temperatures are quite likely to perturb an enzyme, even to the extent of deactivating it. The situation becomes even more complex where a given protein can fulfil two functions: erythrocyte catalase combines catalytic and peroxidase activity in its normal tetrameric state. A thermal shock leads to its dissociation into two dimers, causing a decrease in the catalytic activity but an increase in peroxidase activity.

Douzou lays great stress on adjusting the dielectric permittivity (ϵ) of the cryosolvent to the value of water at the physiological temperature, i.e. 80. The temperatures at which $\epsilon = 80$ for various 50% v/v aqueous mixtures are: methanol -32, dimethyl sulphoxide (Me$_2$SO) $+9$, dimethyl-formamide -21 and ethane diol $-18\ °C$. There is thus a wide range of ϵ/T conditions which can be used to maintain the reaction mixture at $\epsilon = 80$.

Another important criterion is said to be the pH-activity profile of the reaction, the shape of which should be unaffected by the cryosolvent if an unchanged reaction mechanism is postulated. We have already discussed temperature dependent changes in pK_w and pK_a of acids. Such changes are easy to understand and interpret. The significance of hydrogen ion activity and acid dissociation in mixed solvents is not so obvious (Hepler & Woolley, 1973), nor is the correct interpretation of a pH-activity profile in such a solvent mixture (Taylor, 1979).[†]

The features discussed so far would appear to be the minimum conditions which must be satisfied, but they are by no means sufficient proof that the reaction pathway in the cryosolvent corresponds in every detail to that observed in a normal aqueous medium. In fact, such a correspondence is extremely unlikely, bearing in mind the sensitivity of protein conformation to environmental perturbations. For instance, the thermal denaturation transition is shifted, usually to lower temperatures, as organic solvents of the kind used in cryomixtures are added to water (Parodi, Bianchi & Ciferri, 1973). Sometimes such changes are extremely complex as the organic solvent content is increased (Brandts & Hunt, 1967).

It would appear to be quite feasible that by substituting a cryosolvent for water, the (apparent) pH-activity profile of the enzyme will not be markedly affected and the product of the catalysed reaction will be the same as in the conventional solvent medium. It is highly unlikely, though, that the reaction pathways are identical, and even if they are, that the nature and energetics of the two transition states would be identical, depending, as they do, on the solvated state of the enzyme. It would be pure coincidence if the $\Delta G(T)$ relationship of a given protein would be the same

[†] The published pH values of buffers in cryosolvents require comment. They were obtained by means of e.m.f. measurements in which the glass electrode was immersed in the cryosolvent at some subzero temperature, but the reference electrode was immersed in aqueous KCl at *room temperature* (Douzou, 1977). Quite apart from indeterminate liquid junction potentials, these are not the conditions under which *reversible* e.m.f.'s should be measured.

in two so different solvent media (see Fig. 4.3 for the free energy surface of an enzyme as a function of a specific solvent property, pH). One way of testing the hypotheses is to compare the kinetic details of the reaction in the conventional aqueous medium at subzero temperatures with those in a cryosolvent. This requires the inhibition of ice nucleation over as large a range of temperature as possible. The dispersion of the aqueous phase as a droplet emulsion in an inert carrier fluid has already been discussed and has been applied as a test for possible kinetic or mechanistic perturbations caused by cryosolvents.

Complexes between haemoproteins and carbon monoxide can be dissociated photochemically and the recombination rate measured. This has been done for horseradish peroxidase (HRP) at subzero temperatures in undercooled water (Douzou, Balny & Franks, 1978) and in a 50% v/v solution of ethane diol (ED) (Douzou, 1977). The recombination rate constants are shown in the form of Arrhenius plots in Fig. 4.9. As demanded by the criteria set for the validity of cryosolvents, both Arrhenius plots are linear, but the one for undercooled water has the same

Fig. 4.9. Arrhenius plot of the kinetics of combination of horseradish peroxidase and carbon monoxide in undercooled water ($\triangle$) and a 50% ethane diol/water mixture ($\bigcirc$), both at subzero temperatures. Note that the slope of the undercooled water experiment is identical to that observed at ordinary temperatures in aqueous solutions ($\bullet$). Reproduced from Franks (1982c).

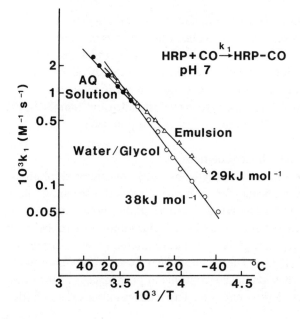

slope as that obtained for a homogeneous aqueous solution at normal temperatures. The activation energy is 29 kJ mol^{-1} compared to 38 kJ mol^{-1} for the cryosolvent. Without additional, thermodynamic information it is impossible to assess whether the ground state, the transition state or both states are affected by the cryosolvent, but the kinetic data clearly indicate a difference in the reaction pathways.

The other reaction for which a comparison has been made between cryosolvents and undercooled water is the luciferase promoted oxidation of reduced flavine mononucleotide (FMN) by molecular oxygen (Douzou, Balny & Franks, 1978). Under *in vivo* conditions the reaction gives rise to bioluminescence (Douzou, 1977). One of the intermediates has been isolated and purified with the aid of a cryosolvent (50% ED); it consists of the enzyme/substrate complex linked to molecular oxygen: $E-FMN-O_2$. At -20 °C this intermediate has a lifetime of several days and can be purified and used as starting material in the study of the bioluminescent oxidation steps. It is found that in the presence of the cryosolvent the reaction proceeds by a 'dark' pathway, whereas in undercooled water the characteristic bioluminescence is observed. On the other hand, the solvent medium does not affect the nature of the reaction products. One can conclude that the criteria proposed for the validity of cryosolvents as reaction media are necessary but not sufficient, because the enzyme energetics are affected by the mixed solvent and this, in turn, is likely to alter the details of the reaction pathway from that normally associated with the *in vivo* process.

Despite the limitations arising from solvent perturbations of enzymes, cryoenzymology in mixed solvents has been shown to be a powerful tool in the elucidation of kinetics, the identification of intermediates and the isolation and purification of labile species. On the other hand, the use of undercooled water in the form of a droplet emulsion, despite some experimental problems, is more likely to provide detailed and unambiguous information about *in vivo* reaction energetics and pathways.

4.4   Proteins in partly frozen solutions

The complex behaviour exhibited by freeze concentrated solutions becomes even more complicated for solutions of natural macromolecules, because of the irreversible processes that can occur during freezing and/or thawing. Freeze induced changes are of particular importance in the food industry where frozen storage is a standard method for ensuring an adequate product shelf life. The technological literature abounds with catalogues of freeze–thaw damage symptoms in proteinaceous substrates (Fishbein & Winkert, 1979), but little basic information exists on the combined effects of concentration and low temperature on proteins in multicomponent systems. In general, proteins are subject to freeze denaturation, resulting at least partly from concentration. In some cases cross linking can occur, followed by insolubilization. The extent of denaturation depends on the initial pH of the solution, the protein concentration, the temperature at which the part frozen solution is kept and the presence of any other substances in solution. The degree of denaturation also depends on the period of exposure to freezing conditions. Figure 4.10 shows the rather complex dependence of chymotrypsinogen denaturation on temperature and time (Brandts, Fu & Nordin, 1970). The rate constant for the reaction increases with decreasing temperature, but a constant degree of denaturation is reached in each case; this equilibrium (or stationary state) value is seen not to bear a simple relationship to the temperature. A quantative analysis of the denaturation curves indicates that the process cannot be described in terms of the usual two state $N \rightarrow D$ transition, where denaturation levels substantially less than 100% would be accounted for by equilibrium constants close to unity. Rather, denaturation is governed by kinetics.

Considering now the denaturation in the time independent range, Fig. 4.11 shows the effect of temperature and additives at an initial pH 1.87 (chymotrypsinogen does not exhibit freeze denaturation at 2 < pH < 10). The sharp maximum suggests that the observed effect is the result of at least two contributions. The dramatic changes produced by low concentrations of salt or nonelectrolyte require comment. Thus, NaCl inhibits denaturation down to −30 °C, at which temperature the salt probably

Fig. 4.10. Kinetics of freeze denaturation of chymotrypsinogen at different temperatures; initial protein concentration 0.25 wt %, pH 1.78 at 25 °C. Reproduced from Brandts, Fu & Nordin (1970).

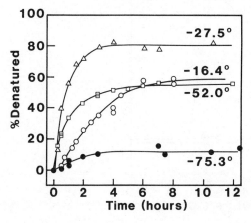

precipitates (in the two component system H₂O–NaCl the eutectic phase separation occurs at $-21\,°C$). The HCl concentration in the unfrozen liquid therefore increases, and the common ion effect will further reduce the solubility of NaCl. Eventually the unfrozen liquid will reach the same composition which it would have in the absence of NaCl.

Sucrose completely inhibits denaturation at all temperatures. In a dilute liquid mixture its effect would not be nearly as pronounced. This effectiveness of sucrose (and most other polyhydroxy compounds) in preventing freeze denaturation cannot yet be fully explained; thermodynamic studies of ternary or quaternary systems containing proteins and sugar alcohols do however indicate a stabilizing effect (see next section). The protective influence of sugars on protein integrity during excessive processing is exploited industrially in freeze drying operations.

Urea is well known as a protein destabilizer (Franks & Eagland, 1975) and its pronounced effect on freeze denaturation is undoubtedly due to its enhanced concentration in the residual liquid. It is curious that the effect is so marked even at 0 °C and also that there always exists a small fraction of undenatured protein. Protein concentration also affects the degree of denaturation of chymotrypsinogen. At $-20\,°C$ and pH 1.81 the per cent denaturation increases to a maximum of 70% at a protein concentration of 10% by weight. It then falls off gradually to 20% at a protein

Fig. 4.11. The effects of additives on the freeze denaturation of chymotrypsinogen: 0.1 M NaCl (▲), 0.1 M sucrose (○) and 0.1 M urea (□). The drawn-out line describes the denaturation behaviour in the absence of additives. Data from Brandts, Fu & Nordin (1970).

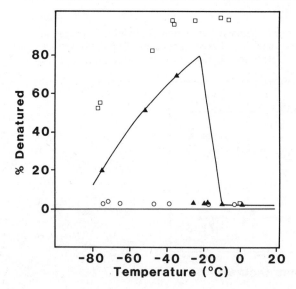

concentration of 75%. The stabilization of proteins by concentration seems to be a general phenomenon. It can also be achieved by the addition of a second protein. Thus, the addition of serum albumin protects enzymes against heat- or pH-induced denaturation. The degree of stability achieved through concentration can be quite marked: when a solution of chymo-trypsinogen is dried to a moisture content of 11%, its thermal denaturation temperature is thereby raised to 100 °C.

Since freeze inactivation of proteins is under kinetic control, the rate of cooling also plays a part in determining the extent of denaturation. With dilute solutions rapid cooling minimizes inactivation. Cooling rate effects are not nearly as marked with concentrated solutions, for reasons already discussed in chapter 3.

As might be expected, the influence of freezing on the kinetics of enzyme catalysed reactions is complex, especially in cellular systems and in systems where the enzyme is membrane bound. Freezing generally causes rate enhancement of enzyme catalysed reactions in cellular systems, whereas in noncellular systems such enhancement is only observed where the initial concentrations of the reactants (prior to freezing) were extremely low (Fennema, 1975). Since the kinetics of enzyme reactions at subzero temperatures are of great importance in food processing and storage, most of the available studies have concentrated on such systems. In general, the reaction rate/temperature relationships resemble that shown in Fig. 3.14 for the simple mutarotation of glucose under conditions of freeze concen-tration. Thus, the rate of hydrolysis of phospholipids in fish rises to a maximum at −4 °C, as does also the rate of ATP depletion in meat muscle tissue. The enhanced loss of vitamin C in partly frozen fruit and vegetable products is of particular concern (Thompson & Fennema, 1971). Apart from the direct effects of freeze concentration it appears that free radicals contribute to the observed loss: the stability of free radicals which is low in systems of high water content is considerably increased under low moisture and subfreezing conditions.

Rate enhancement in noncellular enzyme systems at low temperatures is rare; examples include the decomposition of peroxides by catalase and the oxidation of guaiacol by peroxidases when frozen slowly. In most published reports the results refer to assays performed after the reaction mixture has been thawed. This practice must always introduce a degree of uncertainty, because the observed rate enhancement may have occurred during thawing. In some cases this has in fact been established by comparing the assays in the frozen state with those after subsequent thawing (Fennema, 1975).

The high reaction rates in frozen cellular systems are probably a direct consequence of the disruptive effects of freezing on membranes. This will be more fully discussed in the following chapter. Here we shall limit ourselves to the freeze induced kinetics and inactivation phenomena. The best documented example is the effect of freezing on the functioning of thylakoids (Heber *et al.*, 1981). Their ability to catalyse the synthesis of ATP is much reduced by high concentrations of NaCl. Freezing further sensitizes the membranes towards the effects of NaCl, as shown in Fig. 4.12: the NaCl concentration required to produce a 50% loss of photophosphorylation capacity decreases from 0.44 M at 0 °C to 0.05 M at −12 °C. However, bearing in mind the amount of freeze concentration at the lower temperature, the actual NaCl concentration in the unfrozen liquid is 1.4 M.

The cause of thylakoid inactivation is not certain; the symptoms of *in vivo* injury do not appear to be the same as those found in isolated membrane systems. In any case, damage increases during thawing. The role of specific toxic solutes, which might accumulate during freezing *in vivo*, cannot be ruled out. A much more general effect would seem to be the partial disintegration of the membrane structure, leading to a delocalization of membrane bound proteins.

4.5   The protection of proteins against denaturation

Reference has already been made to the ability of some salts and nonelectrolytes to stabilize native proteins against denaturation. The effects of cosolvents in the stabilization of proteins against thermal

Fig. 4.12. The influence of temperature on the salt-induced inactivation of thylakoids. Initial NaCl concentrations are shown; freeze concentration has not been allowed for. After Heber *et al.* 1981).

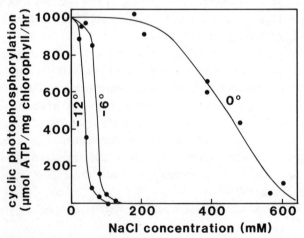

denaturation have received most attention, and Table 4.2 summarizes the available information. It appears that some salts are able to protect proteins against the effects of high temperatures, whereas others destabilize the native state. The same is true for organic nonelectrolytes: as a general rule alkanols destabilize, whereas polyhydroxy compounds, such as sugars or sugar alcohols, stabilize native proteins. The phenomenology of such effects is now well established (Arakawa & Timasheff, 1982), but the causes are still quite unclear. The order in which ions perturb native states is not confined to proteins but has also been observed in the way in which they affect the solubility of argon (and other gases) in water. Thus, the ions that destabilize native proteins enhance the solubility of argon, while those ions that stabilize proteins against thermal denaturation reduce the solubility of argon. The order of effectiveness of the ions is referred to as the Hofmeister series (Hofmeister discovered the phenomenon in 1888) or the lyotropic series. Although the series has been studied in detail (von Hippel & Hamabata, 1973), and finds application in many other areas of science and technology, there is as yet no theoretical basis that might explain the observed effects. The same is also true for the influence of organic molecules on protein solubility and conformation. Here again, the observed effects can hardly be explained in terms of specific interactions between proteins and the substances involved, because these same substances act in a similar manner in much simpler systems, such as argon in water. Indeed, the only common factor is water; it seems likely, therefore, that

Table 4.2. *Effect of solutes on the thermal unfolding temperature of ribonuclease,* ΔT_m *(deg mol^{-1})*

Additive	pH	ΔT_m
KH_2PO_4	6.6	+16
$(NH_4)_2SO_4$	7.0	+11
KCl	7.0	−1
LiCl	7.0	0
NaBr	7.0	−1
LiBr	7.0	−6
$CaCl_2$	7.0	−9
KCNS	7.0	−15
$(CH_3)_4NBr$	7.0	−5
$(C_4H_9)_4NBr$	7.0	−36 (extrapolated value)
CH_3OH	7.0	−1
C_3H_7OH	7.0	−7
Glycerol	4.7	+1
Erythritol	4.7	+2
Sorbitol	4.7	+4
Urea	7.0	−4

any explanation of the experimentally observed effects should be sought in terms of interactions between hydration spheres of different kinds.

Although the exact molecular mechanism of the effects described is still shrouded in uncertainty, it is nevertheless possible to obtain a quantitative thermodynamic picture of the protective or destabilizing influence of different kinds of solutes on protein integrity. The quantitative description of three or multicomponent systems is somewhat complex. For the Gibbs–Duhem equation to be applied, measurements must be made under each of the following conditions, where subscript 2 refers to protein and 3 to the additive: change the concentration of 2, keeping that of 3 constant, change the concentration of 3, keeping that of 2 constant, and changing the concentration of both 2 and 3, but keeping the corresponding chemical potentials μ_2 and μ_3 constant (Eisenberg, 1976).

The basic equation relating a measured property X to the various concentrations m_2 and m_3, at constant temperature and pressure, can be written as (Arakawa & Timasheff, 1982)

$$\left(\frac{\partial m_3}{\partial m_2}\right)_{\mu_1,\mu_3} = \frac{(\partial X/\partial m_2)_{\mu_3}-(\partial X/\partial m_2)_{m_3}}{(\partial X/\partial m_3)_{m_2}} \qquad (4.5)$$
$$= -(\partial\mu_3/\partial m_2)_{m_3}/(\partial\mu_3/\partial m_3)_{m_2}$$

so that

$$(\partial\mu_2/\partial m_3)_{m_2} = (\partial\mu_3/\partial m_2)_{m_3} - -(\partial m_3/\partial m_2)_{\mu_1,\mu_3}(\partial\mu_3/\partial m_3)_{m_2}$$

from which, as a good approximation,

$$(\partial\mu_2/\partial m_3)_{m_2} = -(\partial m_3/\partial m_2)_{\mu_1,\mu_3}(RT/m_3) \qquad (4.6)$$

Equation (4.6) indicates the influence of added substance 3 on the thermodynamic stability of the protein in solution.

Also, for component 1 (water):

$$(\partial m_1/\partial m_2)_{\mu_1,\mu_3} = -(m_1/m_3)(\partial m_3/\partial m_2)_{\mu_1,\mu_3} \qquad (4.7)$$

The property X might be the density, vapour pressure, specific heat or some spectroscopic parameter. The unit of concentration chosen in eqns (4.5), (4.6) and (4.7) is molality (mol in 1 kg of water). The left hand side terms in eqns (4.5) and (4.7) are the preferential interaction parameters (Eisenberg, 1976) which indicate whether in the proximity of the protein there is an excess or a deficit of component 3 or 1 respectively. Thus, for $(\partial m_3/\partial m_2) < 0$ there will be a deficit of species 3 in the immediate neighbourhood of the protein. In other words, there will be an excess of water, over and above that quantity corresponding to the bulk composition of the solvent mixture. The concentration excess or deficit has a statistical significance only and it is not helpful to describe the effects in terms of binding of the solvent components to the protein.

Table 4.3 summarizes the preferential interaction parameters for bovine serum albumin (BSA) and a series of solutes in 1 M aqueous solution.

Negative values indicate exclusion of solute from the proximity of the protein, implying that there is a hydration layer which is not readily penetrated by these solutes. In thermodynamic terms this means that the protein in solution is destabilized, i.e. its chemical potential is increased, as shown by eqn (4.6). Considering now the relative influences of the various solutes on the stabilities of the N and D states, the experimental results indicate that the N state suffers less destabilization than does the D state, possibly because of the larger solvent exposed surface area of the latter. The net effect is therefore a protection of the native state which shows itself as a rise in the temperature of thermal denaturation (Gekko & Morikawa, 1981).

It was earlier pointed out that cold denaturation (or dissociation) exhibits the same thermodynamic trends as does thermal denaturation. It would therefore be expected that those solutes which protect the N state would do so equally at low temperatures, and since the effect seems to be one of maintaining the hydration shell of the protein under conditions of lowered water activity, the protective effect should also be observed during the freeze concentration of the protein. The recognition of such effects is reflected in the application of sugars during industrial spray and freeze drying operations. Rather more remarkable is the *in vivo* synthesis of polyhydroxy compounds in response to physiological water stress conditions, whether caused by salinity, drought or freezing. Although such compounds may also fulfil other protective functions, their role as *in vivo* protein stabilizing agents cannot be ruled out.

Table 4.3. *Preferential interaction parameters at 25 °C of bovine serum albumin with cosolutes in mixed aqueous solutions, according to eqn* (4.5)

Units: mol cosolute/mol protein

Cosolute	Concentration	$-(\partial m_3/\partial m_2)_{\mu_1,\mu_3}$
Ethane diol	20% (v/v)	50
Glycerol	20% (v/v)	38
Xylitol	20% (w/v)	13
Mannitol	15% (w/v)	13
Sorbitol	15% (w/v)	12
Inositol	10% (w/v)	15
Na_2SO_4 (pH 4.5)	1 molal	32
2-Chloroethanol	40% (v/v)	−482
Glycine	2 molal	62
Lactose	0.4 molal	10
Glucose (pH 6)	0.5 molal	11
NaCl (pH 4.5)	1 molal	18
$CaCl_2$ (pH 5.6)	1 molal	−2
KCNS (pH 5.6)	1 molal	−5

5

The single cell: responses to chill and freezing

5.1 The cell as a heterogeneous system

The previous chapters have dealt with the physical and chemical aspects of subzero temperatures with emphasis on differences between undercooling and freezing. It became apparent that undercooling does not affect the homogeneity of a system, whereas the prime consequence of freezing is the creation of a heterogeneous, disperse system, following the phase separation of ice as a pure water phase. We now turn to an examination of the influence of low temperatures and freezing on living cells. In this context the most important feature is the physical and chemical asymmetry which is required and must be maintained for proper physiological functioning of even the simplest of biological systems. Thus, at the most basic level, the chemical composition of the cell interior differs markedly from that of the extracellular medium, although the two domains are in osmotic equilibrium. They are, however, also in communication, with material and energy being constantly exchanged between the cell and its surroundings. The cell membrane functions as a mediator for such exchanges, and it is therefore not surprising that malfunctions in the membrane have been identified as prime causes of chill and freeze injury in cells (Morris & Clarke, 1981).

What is true for the relationship between the cell and its surroundings is equally true for the intracellular space which contains several different kinds of subcellular structures and organelles suspended in the cytoplasm. Here again each unit of organization is surrounded by a membrane which defines its spatial domain and maintains its chemical composition. Some organelles and structures are universal, or almost so; they include the nucleus, the ribosome and the mitochondrion. Others are more specialized, e.g. the chloroplast and the tonoplast in plant cells.

At an even higher level of resolution, the chemical composition of cells is non-uniform, with each organelle or other supramolecular structure maintaining its own heterogeneity. Thus, the distributions of the simple ions Na^+, K^+, Mg^{2+} and Ca^{2+} vary at different locations within the cell, and specific physiological functions are associated with temporal changes in these distributions. The maintenance of ion concentration gradients requires energy which, in living organisms, is derived from the hydrolysis of ATP. A prime requirement for survival at suboptimal temperatures must therefore be an adequate supply of ATP and an adequate rate of ATP conversion.

Finally, in this brief survey of cell heterogeneity, attention is drawn to the molecular details of the various membrane systems which are so vital to the transport between organelles and between the cell and its surroundings. The basic structural unit of the vast majority of membranes is the phospholipid bilayer. Singer & Nicholson (1972) first proposed the currently popular model of the 'fluid mosaic', according to which considerable diffusional freedom exists within the plane of each layer.

The phase and dynamic properties of such phospholipid–water systems have been extensively studied and are found to be extremely sensitive to (1) the distribution of fatty acid chains, (2) their degree of unsaturation, (3) the nature of the polar headgroup, in particular whether headgroups bear an ionic charge, (4) the chemical dissimilarity between the interior (facing the cell) and exterior (facing the extracellular medium) layers of the membrane and (5) the nature of other, nonlipid constituents held within the phospholipid bilayer matrix, in particular cholesterol, peptides and glycopeptides. Since even pure phospholipids dispersed in water exhibit marked structural and dynamic changes with changes in temperature (thermotropism), it is not surprising that many investigators associate the membrane with the incidence of low temperature injury in living organisms.

5.2 Low temperature stress – biochemical aspects

The concept of physiological stress has already been briefly mentioned; here we deal with temperature as the stress factor. Reference to the scientific literature shows that high temperature receives much more attention than do subambient temperatures (Precht, Christophersen, Hensel & Larcher, 1973), and where low temperatures are studied at all, in many cases no clear distinction is made between the effects of the temperature *per se* and those due to freezing.

The chemical processes which contribute to the wellbeing of the cell are so diverse and interrelated that it would be difficult, if not impossible, to calculate analytically the overall effect of temperature on growth and reproduction. Taking even a single biochemical function, protein synthesis, as a measure of overall physiological functioning presents insuperable problems, since even simple microorganisms, such as *E. coli*, contain upwards of 1000 different proteins. It cannot be assumed that the rates of synthesis of different proteins are identically affected by changes in temperature. Nevertheless, most available data on temperature effects on biological processes are interpreted in terms of simple Arrhenius plots of log (rate constant) *vs.* $1/T$. Linearity of such a plot is often taken as an indication that the overall process under study (e.g. the growth of a cell culture) follows the simple kinetic relationships that govern isolated chemical reactions. The simple explanation found in standard text books is that the rate of the total process is governed by the slowest reaction step in a sequence. Attractice though this simple interpretation may be, it can hardly be upheld in the face of thorough investigations, analytical and statistical, into the significance and applicability of the Arrhenius relationship (Krug, Hunter & Grieger, 1976; Livingstone, Franks & Aspinall, 1977; McMurdo & Wilson, 1980; Bagnall & Wolfe, 1982). While it is inappropriate to discuss the results of these studies in detail, some of the results have a very direct bearing on physiological speculations about origins of chill and freeze injury in living organisms.

Most biological rate studies can only be performed over very limited ranges of temperature, typically between 15 and 35 °C. While the measured rate constant may change by orders of magnitude, the temperature interval corresponds to only a 6% change in $1/T$ of the value corresponding to 15 °C (2.3×10^{-4} in 3.47×10^{-3} K^{-1}). The assumption of linearity and the assignment of an energy of activation ($\Delta E^{\ddagger}$) are highly questionable. For simple organic reactions in aqueous solution, extending over much wider temperature ranges than are possible for biological processes, it has been shown that the Arrhenius plots are not linear (Robertson & Sugamori, 1972; Engberts, 1979). Indeed, knowing what we do about water as a solvent, it would be surprising for the energy of activation to be constant, independent of temperature. Careful statistical analyses of kinetic data indicate that for any quantitative significance to be attached to an Arrhenius plot requires rate constant measurements of a quality such that the uncertainty in log k does not exceed 0.1% (Krug, Hunter & Grieger, 1976), a requirement that can hardly be met in biokinetic studies.

Another serious shortcoming of the Arrhenius-type processing of data lies in the estimation of entropies, as well as energies (enthalpies) of activation. In other words, the kinetic measurements are made to yield the intercept as well as the slope of the plot. This involves an extrapolation of the plot to $1/T = 0$ from a series of experimental data extending over only 2.3×10^{-4} K^{-1}. Once again, there is a high statistical correlation between the errors in the slope and the intercept, and data of the highest precision are required for the estimation of significant $\Delta S^{\ddagger}$ and $\Delta E^{\ddagger}$ values. It is unfortunate that numbers obtained from cell growth experiments, or even from the more simple enzyme kinetic studies, do not even come close to this demanding criterion and cannot therefore provide quantitatively significant information about molecular processes.

Bearing in mind the severe limitations of Arrhenius plots as quantitative measures of biological processes, let us examine nevertheless what information has been provided by kinetic measurements. The growth of a cell culture can be divided into several stages: (a) a lag phase during which the number of cells hardly shows any change, (b) the accelerated growth phase when an upturn takes place in the total number of cells, (c) an exponential (log) phase, during which exponential growth is observed, (d) a decelerated phase during which the rate of growth decreases, eventually to reach (e) the stationary phase during which the cell number remains constant. Finally the cells die, once again by an exponential decay, ending in a stationary phase.

The three major growth stages are (a), (c) and (e), each of which is subject to a temperature dependence. Quite apart from the influence of temperature on the cellular processes, corresponding effects take place in the cell environment and thus produce secondary changes in biosynthesis, reproduction and metabolism. These secondary effects are often neglected in measurements of cell growth at different temperatures; they were discussed in earlier chapters of this book. The exponential growth phase is the easiest one to study because it is easiest to handle mathematically. Figure 5.1 shows a typical growth curve/temperature relationship; the inset is the Arrhenius representation of the data. The minimum corresponds to the optimum temperature for reproduction, in this case $\sim 40\,°C$ (Buchanan & Fulmer, 1930). However, the curve is not symmetrical about this temperature; it rises quite steeply at higher temperatures, with almost a cut-off at 50 °C, whereas at low temperatures it exhibits a much more gradual rise. For practical purposes most cell types are capable of growth over a temperature range of about 30°, but the optimal growth temperature is nearly always substantially below those where maximum enzyme activities are observed and far below those where enzymes become inactivated. Despite this, temperature induced growth inhibition is generally claimed to be associated with enzyme inactivation.

Several attempts have been made to provide mathematical descriptions of growth curves of microorganisms. At the simplest level it can be assumed that growth as a function of temperature depends on two or more competing processes, e.g. the biosynthesis of cellular chemicals and the inhibition (denaturation) of an essential enzyme. By expressing each of these processes by a rate constant, growth then becomes the sum of two exponential functions. Various elaborations of this treatment are on record, some of them able to fit the heat inactivation data. However, this simple enzyme inactivation model must be questioned, because growth rate changes at suboptimal temperatures show a different pattern from those at superoptimal temperatures. More complex models for heat and cold effects on growth have been proposed in terms of simultaneous and consecutive, reversible and irreversible processes. but they must be regarded as curve fitting exercises rather than means for obtaining a better insight into the molecular events that promote or hinder growth.

On the experimental front most effort goes into studies of the effects of heat on cell growth and survival, probably because of the important industrial implications of microbial heat resistance and heat inactivation.

Fig. 5.1. Temperature dependence of the generation time of *E. coli*. Inset: Arrhenius representation of the growth/temperature relationship. Note the asymmetric nature of the curve, with an abrupt cut-off at high temperatures but a gradual decline at low temperatures.

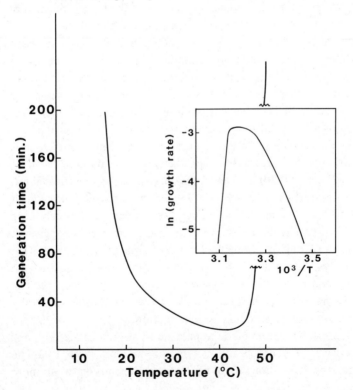

Cold response of microorganisms does not receive much attention; the effects of freezing will be discussed below. Perhaps the most effective method of studying the biochemistry of cold sensitivity/tolerance is by the use of mutants (O'Donovan & Ingraham, 1965). Since a single genetic change alters the optimum growth temperature, the protein modification produced by such a change must be at the basis of the growth/temperature relationship. For instance, cold sensitive histidine mutants from *E. coli* have been compared to the wild types, as shown in Fig. 5.2. The mutant has a minimum temperature of growth which is higher by 12 °C than that of its parent, but when the growth medium is enriched with histidine, the two growth curves become identical. *In vitro* studies of histidine biosynthesis have indicated that cold sensitivity originates from the inactivation of the enzyme in the histidine pathway which is sensitive to feedback inhibition, namely phosphoribosyl adenosine triphosphate pyrophosphorylase (Martin, 1963). Supporting evidence comes from more extensive mutant studies which show that any cold sensitive mutant produces species of this enzyme which are dramatically more sensitive to histidine than that produced by the parent.

The above mechanism of cold inhibition may well be an example of a much more general effect: the temperature sensitivity of allosteric proteins to their effectors. At present it is not possible to predict the effect of temperature on the growth of a given mutant; this would require precise knowledge of the free energy/temperature relationship, as shown in

Fig. 5.2. Growth curves of a cold-sensitive mutant of *E. coli* and its parent. After O'Donovan & Ingraham (1965).

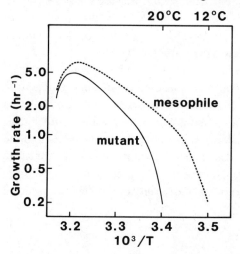

Fig. 4.4 for lysozyme, and the point on the surface which corresponds to optimum growth conditions.

The behaviour of aspartate transcarbamylase derived from different types of cells serves to illustrate this point. The enzyme isolated from *E. coli* is feedback inhibited by cytidine triphosphate (CTP) and activated by ATP. However, the inhibiting effect of CTP decreases with decreasing temperature and disappears completely below 4 °C. On the other hand, the enzyme obtained from *S. cerevisiae* is regulated by uridine triphosphate (UTP), the effect of which becomes increasingly marked with decreasing temperature, reaching a maximum below 3 °C (O'Donovan & Neuhard, 1970; Kaplan, Duphill & Lacroute, 1967).

Other suggested causes of low temperature sensitivity include the production of toxic compounds and the inability of the organism to synthesize ribosomes, possibly due to some minor, but essential, conformational transition in a peptide subunit (Guthrie, Nashimoto & Nomuro, 1979). The *in vitro* assembly of 30s subunits is strongly temperature dependent, with an activation energy of 170 kJ mol^{-1}. If the rate determining step for the *in vivo* assembly is similarly temperature sensitive, and if the driving force depended on hydrophobic interactions, then this would be consistent with the observed cold sensitivity.

Despite many attempts to produce mutants of gram negative bacteria with extended temperature tolerance, the results have been disappointing, from which it is concluded that several mutations would be required. By contrast, aerobic spore formers readily undergo changes in minimum growth temperatures during laboratory cultivation and they adapt to new temperature ranges without difficulty. It is believed that the adaptation is of a physiological rather than a genetic nature, because the properties of the proteins of such organisms depend on the temperature at which they were synthesized, and also because the large majority of cells in a culture can be made to grow at the new temperature.

While an upper temperature limit to growth can be readily rationalized in terms of the limited stability of some cell component, a low temperature limit is not as easily explained. Of the various causes suggested, often based on very tenuous evidence, the two that are consistent with *in vitro* studies of cellular processes are the inability to produce stable organelles and the genetic coding for allosteric proteins. Such interpretations would therefore favour a thermodynamic rather than a kinetic limit to low temperature growth, based on the phenomenon of cold lability of proteins, as discussed in the previous chapter.

5.3   Low temperature effects on membrane processes

At the most basic level the cell membrane provides a barrier between the cytoplasm and the extracellular environment, and its primary function is to maintain the cytoplasmic composition in osmotic, but not chemical, equilibrium with the exterior. In other words, the cell behaves as an osmometer in maintaining a constant volume. By virtue of its mechanical properties it also protects the cell against undue deformation. At a more subtle level the membrane also mediates the selective transfer of material in and out of the cell, and it also contains the various recognition sites for hormones, neurotransmitters and cell–cell interactions. The transport and specific recognition functions are accomplished by proteins which are embedded in the lipid matrix of the membrane and rely for their activities on complex association/dissociation processes stimulated by temperature, pH, electrical impulses, or specific chemical effects. In turn, the ability of the peptides to function optimally depends on the dynamic properties of the lipid matrix, in particular its fluidity. The phase behaviour of phospholipid–water systems is complex and, depending on the lipid:water ratio, the lipids can exist in one of several possible ordered structures in which the individual lipid molecules have more or less motional freedom. For proper *in vivo* functioning of the membrane, i.e. at high water contents, the lipids need to be in the fluid state in which the alkyl groups have a considerable degree of rotational freedom and the lipid molecules are able to perform two-dimensional diffusion in the plane of the membrane. For any given phospholipid–water system there exists a definite temperature at which the lipids adopt a hexagonally packed structure, thus losing their diffusional freedom, and at which the hydrocarbon residues undergo a *gauche* → *trans* transition with the chains becoming fully extended (thermotropism) (Pringle & Chapman, 1981). This array is quasi-crystalline and is referred to as the gel phase; the lipid arrangement above the transition temperature is known as the liquid crystalline phase.

For a chemically homogeneous phospholipid the transition is highly cooperative and resembles a melting process. The actual temperature of the liquid crystal → gel transition depends on the chain length and degree of saturation of the acyl chains, decreasing with increasing unsaturation and decreasing chain length. Thus, for a series of 1,2-diacyl-L-phosphatidylcholines the distearoyl, dipalmitoyl and dioleyl derivatives have transition temperatures of 54, 41 and −22 °C respectively (Chapman, Williams & Ladbroke, 1967). With phospholipid mixtures the transition becomes less well defined and the possibility arises of a two-dimensional phase separation, with each component migrating within the bilayer to produce pure crystalline regions. Such a process can be regarded as a two-dimensional analogue of eutectic phase separation.

Substances which are miscible with phospholipids even in the gel state will tend to suppress the transition. Such a substance is cholesterol, a common ingredient of biomembranes. The effect of cholesterol on the molecular motion of lipid molecules is to inhibit their diffusional freedom at high temperatures but to enhance it at low temperatures. Thermotropism in phospholipids has been studied mainly on artificial membrane systems, phospholipid liposomes. How far the phenomemon has any physiological significance is an open question. For instance, intact erythrocyte membranes do not display thermotropism, but they do so after removal of their cholesterol. Some prokaryotic membranes exhibit transitions extending over a considerable temperature range (weakly cooperative). The transition range can be altered by the addition of specific fatty acids to the growth medium. Alleged discontinuities in Arrhenius plots of cell growth curves have frequently been equated with phase changes in membranes (Lyons & Raison, 1970a, b), although such simple correlations are being subjected to increasingly critical scrutiny (Wolfe, 1978).

Compared to the wealth of information about the physical properties of the lipid membrane components, little is known about the effects of temperature on the functioning of the membrane proteins embedded in the lipid mosaic. Such data on protein distributions as do exist have been derived mainly from freeze–fracture electron microscopy (Fujikawa, 1981). The knowledge that freezing produces gross changes in the distributions of all manner of molecules must cast doubt on straightforward interpretations of changes in the distribution of proteins within the lipid matrix. On the other hand, the direct measurement of protein diffusion is difficult to perform and results appear to differ very much for different proteins and different membranes. The functioning of membrane bound enzymes might also be affected by cytoskeletal (intracellular) components, and it will require much more work with enzymes in reconstituted or synthetic membrane systems before any credible connection can be established between the behaviour of the lipids and the functioning of enzymes in membranes. Where such investigations have already been performed it has been found that there is no simple correlation between the onset of enzyme activity and the thermotropic transition of the lipids in the liposome in which the enzyme was embedded. Indeed, activity has been observed substantially below the lipid transition temperature. In other studies (calcium ATPase in sarcoplasmic reticulum) where a break in the Arrhenius plot had been observed, it was found that the isolated, delipidated and reconstituted enzyme preparation showed a similar break (Dean & Tanford, 1978).

Few membrane transport systems have received as much attention as has the sodium pump in the red cell membrane. It should therefore be a good model for studies of temperature effects. In human erythrocytes the Arrhenius plot exhibits a discontinuity at 22 °C. However, the temperature effect could be due to one of several causes: a direct effect on the pump, changes in the permeability of the membrane to other ions, or the metabolism and availability of ATP (Ellory & Willis, 1981). If the effect is one associated with the pump itself, then questions arise about the immediate environment of the enzyme. It is probable that the micro-environment of the enzyme system does not correspond to the global lipid composition. Dramatic effects on pump activity are produced by an alteration in the cholesterol level, but the reasons for this are unclear. It has also been observed that the membrane permeability to passive diffusion of ions *increases* at low temperatures, a paradoxical result indeed. Efforts to determine which, if any, of the partial enzyme reactions are particularly sensitive to low temperatures have led to the suggestion that it is the conversion of the enzyme from the K^+- to the Na^+-preferring form.

Finally, even more doubt must be cast on the significance of the 22 °C break in the Arrhenius plot, because several other transport systems in the red cell membrane also exhibit such discontinuities, but no two of them at the same temperature (Joiner & Lauf, 1979).

The above discussion demonstrates that the origin of cold inhibition of growth is not yet clear and that neither of the two main hypotheses is entirely satisfactory. That the limit to growth is primarily due to a thermotropic phase transition or a phase separation of lipids must now be considered as most unlikely. The alternative, that the conformation or degree of association of enzymes is critically affected below a certain temperature, appears to be more attractive and corresponds more closely to recent experimental observations. The enzyme which is critical to growth inhibition may then be either membrane bound or soluble. In the latter case one would not expect the inactivation temperature to correspond to any discontinuity in membrane properties.

5.4   Low temperature germination of seeds

From the point of view of this discussion, the chief difference between microorganisms and seeds lies in their respective water contents. Whereas most microorganisms are susceptible to freezing, many seeds can be stored at quite low subzero temperatures without the occurrence of freezing. It is therefore possible to study the effects of temperature on germination right into the subzero temperature regime. The literature on seed germination is profuse and mainly descriptive; it is difficult to extract quantitative information. It is even harder to relate experimental observations to physical or chemical changes which precede and accompany germination. Like growth, germination takes place only over limited temperature ranges, the common behaviour resembling that depicted in Fig. 5.1; i.e. the high temperature limit is almost discontinuous, whereas at suboptimal temperatures the germination rate shows a more gradual decline. The actual temperature for optimal germination depends on factors such as the storage period prior to germination, the storage temperature, the moisture content and various other pretreatments.

Despite the low water content and storage at low humidities, biochemical activity in seeds continues after harvesting; this leads to a gradually increasing temperature tolerance of germination which is shown in Fig. 5.3 (Thompson, 1970).[†] Actual temperatures for maximum germination range from ~ 10 to $\sim 45\,°C$, depending on the species and the method of maturation (Simon, 1981). The phenomenon of chill injury which is common in plants is much rarer in seeds, so that many seeds of chill sensitive plants can be stored at temperatures substantially below those at

Fig. 5.3. Time required for the germination of 50% of a seed sample of *Silene concoidea* as a function of temperature and increasing storage periods. After Simon (1981).

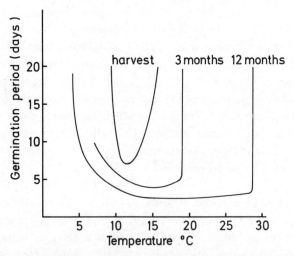

[†] It must be noted that the method of expressing germination rate can be somewhat misleading. By measuring the time taken for 50% of the seed sample to germinate, samples in which < 50% germinate would not be counted at all.

which germination does not occur at all. The so-called recalcitrant seeds are exceptions to this general behaviour: they are damaged by exposure to subgermination temperatures, and also by dehydration (King & Roberts, 1979). The group of recalcitrant seeds includes various tropical species, e.g. mango, avocado, cacao and oil palm. It is open to doubt, however, whether such a classification has any real meaning, because a given seed type would be classified as recalcitrant only until a method for drying and low temperature storage is discovered.

Speculations about possible mechanisms which might limit germination at low temperatures again utilize Arrhenius plots and discontinuities in such plots (Simon, 1979). Although such discontinuities cannot always be correlated directly with the low temperature limit for germination, they indicate temperatures where the energy of activation increases dramatically. Usually the temperature discontinuity lies a few degrees below the limiting temperature for germination. (Note, however, that for germination to be recorded at all, it must be observed for at least 50% of the seed sample.) The explanations offered for the breaks in germination Arrhenius plots follow along the same lines as those for the growth of microoganisms, either in terms of lipid phase transitions and perturbations in membrane-bound enzymes resulting therefrom, or in terms of conformational changes in one or several enzymes, induced by cold.

While microbial growth and seed germination appear to exhibit similar kinetic patterns, such similarities may well be superficial. Seeds, as distinct from microorganisms, contain only $\sim 10\%$ of water, hardly enough to enable the phospholipids to exist in the bilayer state which is characteristic of biological membranes. The available phase diagrams of phospholipid–water systems indicate that, as the water content is reduced to $< 80\%$, the bilayer structure becomes unstable and gives way to a hexagonal lipid phase which would be expected to show a very different thermotropism from that associated with the bilayer structure. A careful search of the literature shows that little is known about the organization of lipids in a resting seed, or about the manner in which normally membrane-bound proteins are organized in the lipid matrix. It does seem unrealistic to interpret low temperature inhibition of germination in such essentially *dry* systems in terms of phenomena observed in, and characteristic of, lipid bilayer structures which probably have no existence in low moisture systems.

As regards the alternative explanation, cold induced changes in protein conformation or assembly of subunits, here again it is commonly found that careful dehydration acts as a powerful protein stabilizing influence, so that in a seed (as also in a bacterial spore) enzymes could be expected to be substantially more resistant towards the effects of heat and cold than they are in fully hydrated (vegetative) systems.

5.5   Cold shock and cold hardening

One parameter that has not yet been discussed in relation to low temperature growth and viability is the rate of cooling. Under natural environmental conditions the cooling rate is not a controllable variable, but in the laboratory, even under simulated field conditions, the rate at which cells are brought to their final nadir temperature may be just as important in determining their survival as the temperature itself. Injury which results from excessively rapid cooling is referred to as cold shock. Like other effects described in this book, rapid cooling and slow cooling are relative terms, so that of two strains of *Amoeba*, one may be sensitive to cooling to $-10\,°C$ at a rate of 0.1 deg min^{-1}, whereas the other may resist cooling rates below 100 deg min^{-1} (McLellan *et al.*, 1984). The quantitative relationship, if one exists, between cooling rate and nadir temperature is not clear, but that the cooling rate does affect low temperature survival has been established for a range of cell types. An alternative manifestation of cold shock is observed with erythrocytes which are normally resistant to cooling, but which become cold sensitive after being subjected to hyperosmotic conditions (Morris *et al.*, 1983).

Evidence of cold shock is illustrated by the data in Table 5.1 which shows comparative survival indices of sycamore cells in suspension after having been cooled rapidly (by immersion at the final temperature) or slowly (at 2 deg min^{-1}) to different final temperatures and maintained there for 5 min prior to rewarming. It is clear that the rapid cooling treatment leads to a significant decrease in the number of cells that survive, whatever the

Table 5.1. *The effects of cooling rate and temperature on the survival of sycamore cells in suspension culture*

Samples were cooled either at 2 deg min^{-1} or by immersion in the cooling bath at the final temperature. Results are expressed as % of the control value (at 25 °C)

Treatment	Temperature (°C)	% survival
Slow cooling	15	97
(2 deg min^{-1})	5	91
	0	88
Fast cooling	15	74
	5	57
	0	66

After Morris *et al.*, 1983.

final temperature (Morris *et al.*, 1983). Qualitatively similar results have been reported for widely differing cell types. Biochemical studies, e.g. protein synthesis, have confirmed the perturbing effects of high cooling rates on cells. Of necessity, the cooling experiments were performed between ambient and high subzero temperatures in order to avoid freezing of the cells.

The question remains whether cold shock is solely a cooling rate effect or whether the final temperature also contributes to the injury. In order to study the latter, a rather different approach has been employed by the author and his colleagues (Franks *et al.*, 1983). Cooling experiments were carried out on cells confined to droplets of aqueous phase emulsified in an inert carrier fluid, in order to prevent inadvertant heterogeneous nucleation of ice. Not only was it found that all the cell types examined could be substantially undercooled (human erythrocytes only freeze at $-38\ °C$), but also that such undercooling did not cause lysis or any other apparent damage. Since the experiments were performed in a scanning calorimeter, the cooling rates could be accurately measured; they were kept in the range $1.25–5\ \mathrm{deg\ min^{-1}}$, which, according to some reports, comes within the range of rapid cooling. No effects of cooling rate could be observed.

In order to account for such divergent results, it has been suggested that in the emulsion type experiments cold injury through material leakage out of the cell might be drastically reduced because of the small extracellular : intracellular volume ratio (Morris *et al.*, 1983). Depending on the cell type these ratios lie in the range from 1:1 and 1:100. In more conventional experiments this ratio would be of the order 300:1. The question of cold shock therefore still awaits resolution. Provided that the conditions which produce cold shock (i.e. fast cooling) are avoided, some cell types are able to develop a degree of tolerance to suboptimal temperatures.[†] The Arrhenius growth profiles of a psychrotrophic *Pseudomonas* and its psychrophilic mutant are shown in Fig. 5.4 (Jaenicke, 1981). The genetic changes which are associated with true psychrophily are unknown, although psychrotrophism can be induced by transduction and mutation. It is not even certain that a general mechanism for cold adaptation exists, because

---

[†] The terminology used in studies of cold tolerance is somewhat confusing. In the following account we distinguish between psychrophilic and psychrotrophic behaviour: the former refers to a genetic adaptation by which the growth/temperature profile is shifted to a lower temperature, whereas the latter refers to a temporary, biochemical tolerance of cold stress. The latter process is also referred to as acclimation or acclimatization, it being understood that, while in this state, the organism in question is not functioning under optimum physiological conditions.

certain enzymes which are thermolabile in one psychrophile may not be so in another psychrophilic organism.

Despite the fact that 80% of the biosphere is exposed either permanently or seasonally to cold (most of the oceans are permanently below 5°C), few truly psychrophilic microorganisms have been identified. Most of the psychrotrophic cells are gram negative bacteria with growth ranges between 1 and 20 °C. Commonly the optimum growth temperature is close to 16 °C. Studies of the ocean bacterial flora have revealed that where the winter temperature is typically −2 °C, psychrophiles predominate, while during the summer, at mean temperatures of 23 °C, mesophiles and psychrotrophs are most common.

Biochemical studies of psychrotrophism have concentrated on the effects of low temperature on the lipid composition of the cell membrane, but the results are equivocal. The membranes of prokaryotes differ from those of eukaryotes in several respects: they contain only trace amounts of sterols, while phosphatidyl serine and phosphatidyl choline are largely absent. The fatty acids show a large predominance of C_{15} to C_{18} straight chain, branch chain or cyclopropane derivatives, saturated and monounsaturated. Di- and triunsaturated acids, which are common in other types of membranes,

Fig. 5.4. Temperature dependence of the specific growth rate of a *Pseudomonas* species and its psychrotolerant mutant. After Jaenicke (1981).

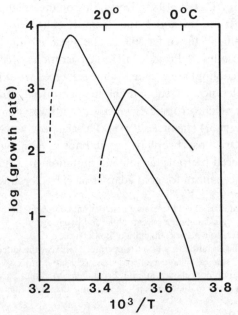

rarely occur (Herbert, 1981). In general the membranes of psychrophiles contain higher proportions of 16:1 and 18:1 acids (palmitoleic and oleic) than do the corresponding mesophiles, and they also respond to low temperatures by synthesizing unsaturated or shorter chain acids. However, this type of response is not universal. Thus, some psychrotrophic marine Pseudomonads, when grown at temperatures lying between 2 and 20 °C, do not produce changes in the fatty acid profile. However, their membrane lipids always contain 67–74% palmitoleic acid.

It is generally held that the plasma membrane lipids of prokaryotes must always be in the fluid liquid crystalline state, i.e. above the gel transition temperature. Possibly this is achieved by the relatively high content of chain branched fatty acids which are rarely found in the membranes of eukaryotes. However, a few cases have been reported of normal cell growth and division under conditions where significant proportions of the membrane lipids are in the ordered state (Jackson & Cronan, 1978; Jarell *et al.*, 1982).

Little is known about the exact mechanisms by which cold triggers the lipid alterations. Possibly desaturase enzymes are induced by cold exposure; on the other hand the observed changes may be brought about through changes in the specific activities of certain enzymes. It is found, for instance, that no synthesis of unsaturated acids takes place in *B. megaterium* above 30 °C; in fact, no desaturating enzyme can be detected. However, on exposure to temperatures below 20 °C palmitic acid is rapidly converted into its monounsaturated derivative (Jaenicke, 1981).

Finally it should be emphasized that an alteration in the lipid composition in response to cold exposure does not necessarily convert the organism (e.g. *E. coli*) into a psychrotroph. Many microorganisms respond to low temperatures by changes in their membrane lipid composition without thereby developing cold tolerance, Indeed, other apparent symptoms of cold tolerance may be equally, or even more, important, such as changes in the turnover of substrates. At present it is not always possible to distinguish between cause and effect in studies of cold tolerance and thermophily.

5.6   Freezing of single cells

At first sight the response of cells to subfreezing temperatures appears to be better understood than does their reaction to low temperature. There are even quantitative treatments to describe the relationships between cell volume, solute concentrations, temperature and rate of cooling/warming. Such relationships are of little relevance to conditions as they exist in the *in vivo* environment, but they become important in the context of cryobiology, the long term laboratory preservation of biological material (see Chapter 8).

Most of the theoretical work and many of the exploratory practical studies have been performed on human erythrocytes, hardly typical of living cells in general. To make the problems involved in the calculations tractable, the erythrocyte is usually regarded as a 'balloon', bounded by the plasma membrane and filled with haemoglobin and electrolyte solution in osmotic equilibrium with the exterior. The membrane is assumed to be permeable only to water, and the intra- and extracellular fluids are assumed to behave as ideal solutions (Silvares, Cravalho, Toscano & Huggins, 1975). Even with such an oversimplified model the calculations are formidable, so that the transport equations have to be solved numerically. In addition, several of the quantities that appear in the mathematical treatment are unknown, while others are difficult to test experimentally. Despite these limitations, the quantitative studies of erythrocyte freezing have helped to provide an insight into the nature and occurrence of freezing injury and have led to the development of improved preservation techniques.

A representation of the model system is given in Fig. 5.5. Early

Fig. 5.5. Simplified model of the kind used for the calculation of cooling rate effects on cell freezing and survival. N is the number of mols of water (w) and solute (s). Superscripts have the following meaning: $'$ = intracellular, i = ice, l = liquid. Water transport can take place between the solid (ice) phase and the extracellular solution phase and between the solution phase and the interior of the cell. The cell membrane is impermeable to all solute molecules.

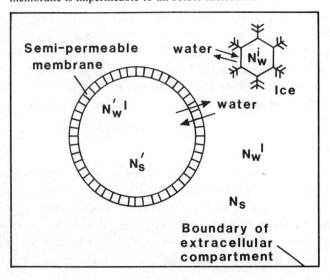

observations indicated that, as the temperature is lowered, freezing first occurs in the extracellular medium, provided that the rate of cooling is low. As the cooling rate is increased beyond a certain threshold value which differs from one cell type to another, so ice is observed to form inside the cell (Mazur, 1970). In either case freeze injury results from increases in the solute concentration that accompany the separation of ice in the system. Under conditions of slow cooling, with ice forming in the extracellular spaces, the extracellular solute concentration increases and the osmotic equilibrium between the cell and its surroundings is disturbed. Water flows out of the cell until the equilibrium is reestablished. The water flux is governed by the permeability of the membrane; at low cooling rates or in a system that is seeded with an ice crystal at the equilibrium freezing point (i.e. under conditions of zero undercooling), osmotic equilibrium is maintained. On the other hand, during rapid cooling, the flow of water from the cell is not fast enough to prevent substantial undercooling of the cytoplasm which eventually leads to intracellular freezing and the accompanying rapid freeze concentration of the cytoplasm. Since both the above cooling regimes lead to freeze concentration of the cell contents, they are both injurious. The retardation of chemical reactions (including the injurious ones) partly balances these effects, so that the relationships governing the degree of freeze damage as a function of cooling rate are complex.

The analysis of the freezing process is performed by considering four distinct stages: (1) cooling prior to the nucleation of extracellular ice; (2) cooling below the nucleation temperature of ice, but while an extracellular liquid phase still exists, i.e. above the eutectic temperature of the extracellular medium; (3) cooling below the eutectic temperature, where the extracellular phase is solid; and (4) holding the system at its low temperature limit until the transfer of water from the cell to its surroundings is complete.

During stage 1 the system is homogeneous and the cell is in osmotic equilibrium with its surroundings. Freezing can occur at, or below, T_f^0, the equilibrium freezing point of the medium, the degree of undercooling to, say, T_{het} being determined by whatever catalytic sites exist for heterogeneous nucleation. At this point the latent heat of crystallization is released and the system warms up spontaneously. The actual trajectory of the broken line in Fig. 5.6 depends on the rate of cooling and the thermal characteristics of the medium. Since the cell is not able to respond instantaneously to the osmotic pressure difference that develops across the membrane upon nucleation, the concentration of water in the cytoplasm will correspond to x_1. During the reestablishment of thermal equilibrium at temperature

T_{eq} the concentration of water in the system decreases to x_2. The loss of extracellular water during freezing leads to a difference in the chemical potentials of water inside and outside the cell. The flux of water across the membrane attempts to reestablish the equilibrium, but it is limited by the permeability of the membrane. As the temperature is lowered, the concentration of water decreases and follows the phase coexistence curve until the system reaches the eutectic temperature T_e. The extracellular medium is now completely solidified, but the physical state of the intracellular fluid and its composition depend on the cooling rate. At low cooling rates the efflux of water from the cell is sufficiently fast for the composition to follow the equilibrium composition curve until T_e is reached and the cell interior will then solidify at that temperature. At high cooling rates, however, the system is far removed from equilibrium conditions when T_e is reached. Undercooled intracellular water will then continue to cross the membrane and immediately freeze. The water contents of model red cells are shown as functions of the cooling rates in Fig. 5.7. It is seen that for cooling rates below 3000 deg min^{-1} the system reaches equilibrium before complete solidification at T_e is achieved.

Stage 4 corresponds to a water transfer across the membrane under isothermal conditions; this only becomes relevant at high cooling rates

Fig. 5.6. Concentration changes during cell freezing in a pseudo-binary system (e.g. NaCl–water). The bold line represents the temperature/concentration profile, with the broken lines indicating changes that depend on the cooling rate.

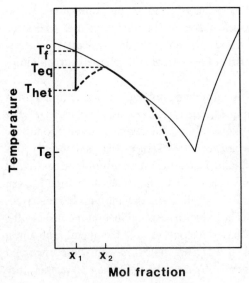

Mol fraction

when the system is far removed from osmotic equilibrium as the limiting low temperature is reached. One further limit to water flux at low subzero temperatures must be considered: as the concentration of solutes in the cytoplasm increases, so its viscosity increases, thus retarding the diffusion of water, quite irrespective of the permeability of the membrane. If the glass transition of the particular cytoplasmic composition is reached, then water flow will be completely inhibited and the system will not be able to reach osmotic equilibrium. The cooling rates required for the vitrification of fairly dilute aqueous solutions are very high indeed and could be achieved only for small samples such as are used in the preparation of specimens for electron microscopy. A further discussion of vitrification of aqueous solutions will therefore be deferred to Chapter 9.

Experience has shown that under conditions of rapid cooling intracellular ice will form. It is not certain whether this occurs endogenously because of the high degree of undercooling or whether it occurs as a result of seeding by extracellular ice. The latter alternative would be expected to apply, bearing in mind that the cell membrane is the only barrier between the frozen phase and the substantially undercooled cytoplasm. The droplet emulsion technique is particularly well suited for the resolution of this question, because cells suspended in the emulsified water droplets can be

Fig. 5.7. Loci of states of intracellular solutions for cells cooled at different rates and in the absence of extracellular undercooling; the composition of the extracellular solution coincides with that given by the equilibrium phase diagram (shown in bold). Data from Silvares *et al.* (1975).

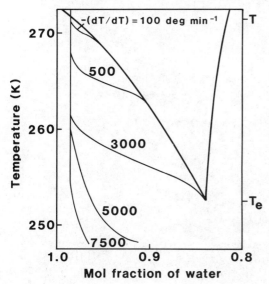

cooled in the absence of extracellular ice, at least down to the homogeneous nucleation temperature of the extracellular medium. The appearance of cells so treated is shown in Fig. 5.8 (Franks *et al.*, 1983). The thin water film surrounding the cells is well revealed at the junction of two cells. The cooling studies have demonstated that many cell types are able to undercool to substantial extents, but that they always freeze *above* the homogeneous nucleation temperature of the extracellular medium; this indicates that the nucleation of ice in cells is always by a heterogeneous mechanism, presumably catalysed by some supermolecular structure within the cell, e.g. the plasma membrane or an organelle. The results from such undercooling experiments indicate that under conventional conditions, cell freezing results from seeding by the extracellular ice.

It is not certain how realistic the simplified model of an erythrocyte might be. Widely divergent results for the residual volume of intracellular water at different temperatures but at the same cooling rate are obtained by only slightly changing the characteristics of the model cells. Several parameters are not included in the model at all, such as the ability of the membrane to accommodate the decrease in cell volume with decreasing temperature, or the fact that there is no *chemical* equilibrium between the intra- and

Fig. 5.8. *Glycine max* cells suspended in emulsified water droplets in a silicone oil carrier fluid. Reproduced from Franks *et al.* (1983).

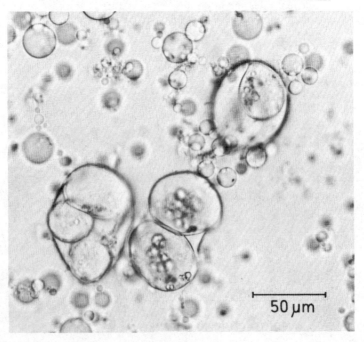

50 μm

extracellular spaces. In addition, the presence of proteins and their properties are not taken into account, nor is the impact of active ion transport on the osmotic equilibrium. Perhaps a better model for freezing studies might be the synthetic phospholipid membrane (vesicle). To make the model more realistic, erythrocyte lipids can be used in the preparation of such vesicles. It has been shown that synthetic membrane systems show close similarity to biological membranes in their responses to freezing (Morris, 1981). Thus, at low cooling rates osmotic dehydration occurs. The rate of water flow can be altered by changes in the lipid composition which is also known to alter the membrane permeability. Damage to the membrane can be monitored by the leakage of marker molecules.

This latter method also reveals differences between natural and reconstituted membranes, as illustrated in Fig. 5.9. The leakage of glucose from synthetic membranes prepared from total erythrocyte lipid extracts differs both quantitatively and qualitatively from the leakage of haemoglobin from intact erythrocytes. Cryomicroscopic studies on synthetic membrane systems show promise of a better insight into the mechanical aspects of freezing stress and injury, and when combined with flux and biochemical measurements, they may eventually yield the complete picture of cold injury and its avoidance.

Fig. 5.9. Glucose leakage from phospholipid liposomes (l.h. ordinate) and haemolysis of erythrocytes (r.h. ordinate) following cooling at 0.25 deg min^{-1} to different temperatures. After Morris (1981).

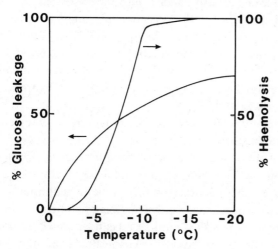

6

Freeze avoidance in living organisms

6.1   Strategies for cold resistance

The alternative strategies for cold survival were briefly referred to in Chapter 1. We have also seen that cold stresses in the absence of freezing must be distinguished from those that occur during freeze dehydration and that the crystallization of ice inside a cell is invariably lethal. For an organism to survive seasonal exposure to subzero temperatures, it must take the necessary measures to prevent its cell water from freezing. In principle, this can be achieved by one of two mechanisms. In the first case, extracellular freezing causes the cell to become partially dehydrated and therefore less liable to freezing. However, the concentration of solutes, especially salts, so produced may also lead to injury and death. Salt injury is avoided through the synthesis of cryoprotective substances. Not only do such substances further depress the freezing point of water, but they may also protect proteins against denaturation, as was discussed in Chapter 4. There is also evidence that biosynthetic cryoprotectants can inhibit lipid phase transitions which might occur during the dehydration of the cell (Crowe & Crowe, 1981).

The other mechanism of cold survival would require a complete avoidance of ice crystallization, at least down to the environmental temperature. In other words, the nucleation or growth of ice within the organism must be completely inhibited, thus allowing the water to undercool. *In vivo* this is achieved either by the synthesis of substantial concentrations of solutes with reliance on their freezing point depressing effects, or by the synthesis of so-called antifreeze peptides which do not affect the *equilibrium* freezing temperature of the tissue fluids but promote extensive undercooling. Figure 6.1 summarizes the alternative strategies used by living organisms to survive periods of subzero temperatures. Apart from those organisms that exist in permanently subfreezing environments (e.g. in the Antarctic ocean), cold resistance requires a period of hardening[†] which, at the end of the cold season will necessitate a period of dehardening. In this chapter we treat the phenomenon of cold resistance by the mechanism of freeze avoidance. The subject can be discussed in connection with the more general topic of the responses of living organisms to low, but non-freezing temperatures, since freeze avoidance only extends the realm of 'chill' into the temperature range where, under normal circumstances, water would freeze. We shall therefore discuss low temperature stress and hardening in general but shall deal separately with the rather specialized mechanisms whereby certain organisms are able to prevent ice from nucleating in their tissues.

6.2   Chill sensitivity and resistance in plants

It is not easy to define chill in quantitative terms, nor is there a very clear borderline between chill sensitivity and resistance. In fact, the definition of freeze resistance is much more clear cut. Perhaps the most useful way to describe chilling is in terms of the changes produced in a plant when it is grown close to its limit of tolerance (Simon, 1981). For

Fig. 6.1. Cold survival strategies. After Levitt (1980).

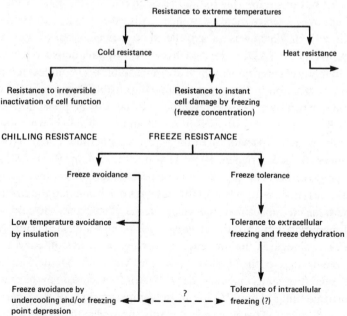

---

† In the American literature seasonal hardening is usually referred to as acclimation.

tropical or subtropical species, e.g. rice, maize or citrus, these limits are fairly clearly defined. We are not concerned here with the visible evidence of cold stress but with changes induced at the molecular level in response to the stress; some of these produce the symptoms of injury, while others produce resistance and enhance the ability of the organism to maintain itself under suboptimal conditions of temperature. These latter processes are those collectively referred to as hardening (or acclimation), and they are best studied by maintaining the plant at a temperature where growth still takes place without visible damage (Graham & Patterson, 1982).

Changes, such as those in enzyme activity or membrane lipid composition, would be expected to parallel those discussed in the previous chapter. During the period of cold hardening changes take place in the protein profile and the activities of many enzymes, but these changes do not solely depend on the temperature but on the light intensity and the photoperiod. Studies on the combined effects of temperature and light intensity on the photosynthetic enzymes in maize leaves have shown that the normal activity of the enzymes is unaltered at 10 °C and 170 w m^{-2} (Taylor, Slack and McPherson, 1974). However, the light activated enzymes NADP-malate dehydrogenase (MDH) and pyruvate phosphate dikinase (PPD) suffer a reduction in activity when the light intensity is reduced to 50 w m^{-2} at 10 °C, but this is not due to a decrease in the enzyme activity but to a loss of enzyme. A sharp increase in the activation energy of PPD was observed below 12 °C which was not paralleled by the behaviour of other enzymes, e.g. MDH or NADP-glyceraldehyde 3-phosphate dehydrogenase.

In many cases enzymes which are quite stable *in vitro* at low temperatures do not function in the intact organism, possibly because of the effects of cold on turnover rates. Thus, the cold lability of phosphofructokinase (PFK) has been investigated in detail and has been shown to arise from a cold induced dissociation of the active tetramer into inactive (but native) dimers (Bock & Frieden, 1976; Dixon, Franks & ap Rees, 1981). Similar *in vitro* dissociation processes have been observed in other multisubunit enzymes (Bock & Frieden, 1978), but it is not certain that the enzymes show the same cold inactivation *in vivo*. On the other hand, the cold sensitivity of mammalian microtubules *in vivo* has been established (Hart & Sabnis, 1977). As regards the influence of low temperatures on enzyme kinetics, it seems that both V and K_m in the Michaelis–Menten equation are affected. For many enzymes the Q_{10} factor (the increase in the reaction rate corresponding to a 10 °C rise in the temperature) is of the order of 2 (i.e. V_{max} doubles). There are some plants, however, for which Q_{10} *increases* at low temperatures! It has been suggested that some phase change(s) in the membrane lipids might be responsible for such an anomalous behaviour which must be regarded as a physiological malfunction due to the low temperature stress (Lyons, Raison & Steponkus, 1979).

As discussed for the case of microorganisms, Arrhenius plots of enzyme processes in plants also exhibit deviations from linearity, but the possibility that K_m may also change with temperature is not often considered, which is surprising in view of the commonly observed curvilinear $\Delta G(T)$ relationship. As was discussed in Chapter 4, the cold inactivation of soluble enzymes is usually associated with the weakening of hydrophobic interactions which are believed to play an important role in the maintenance of multisubunit structures. Where cold lability has been observed in enzymes that are located in membranes, the question again arises whether the inactivation is intrinsic to the protein or caused by lipid phase changes, either of a thermotropic or a phase segregation nature. If the physical state of the lipids is responsible for the observed enzyme changes, then this would be hard to reconcile with the observation that there are no significant differences in the lipid compositions of chill sensitive and chill resistant plants. It must however be borne in mind that very few investigations into the lipid compositions of purified plasma membranes have been performed. The common procedure involves either whole cell extracts of which the plasmalemma comprises 2–4%, or plasmalemma preparations heavily contaminated with other membrane fractions.

Low temperature changes in which the membrane is obviously implicated include discontinuities in the uptake of ions by roots and the leakage of solutes when organisms are suddenly exposed to temperatures substantially below their normal growth temperatures. For instance the blue-green alga *Anacystis nidulans* immediately loses amino acids and ions when it is exposed to a 20 deg chill (Ono & Murata, 1981). Even here, however, the role of the membrane is not quite certain, since chill might well interfere with the production of ATP and/or its utilization.

The process of hardening is a gradual one and appears to be associated with the synthesis of protective compounds. Thus, wheat and potato show the symptoms of hardening at temperatures below 12 °C, while woody species require subzero temperatures for complete hardening (Sakai, 1974). Growth stops during hardening, and the view has been advanced that inhibition of growth is a symptom of injury. It seems that photosynthesis is always maximized at the plant habitat's normal temperature, so that hardening cannot shift the maximum photosynthesis rate to a lower temperature.

Many chemical changes have been identified which are associated with cold hardening. Thus, substantial amounts of sugars and sugar alcohols are commonly produced through the hydrolysis of starch. High levels of proline are found in hardened winter rape, and many plants convert a substantial fraction of the membrane phospholipids into glycolipids (Critchley, 1976). Major changes in the fatty acid profiles have already been referred to. For instance, synchronous cultures of *Chlorella ellipsoidea*, when hardened at 3 °C for 48 hours, produce large amounts of lipids of palmitic, oleic and linoleic acids. The major site of lipid synthesis appears to shift from chloroplasts to a cytoplasmic system (Hatano & Kabata, 1981).

Several distinct stages have been identified in the general hardening process. For instance, on exposure of potato to chill temperatures the first noticeable change is the accumulation of sugars, leading to an increase in the osmotic pressure. Whether this is relevant to the hardening process is not known. Sugar accumulation is followed by release (or synthesis?) of increased levels of abscisic acid which, in turn, is followed by an increase in the rate of protein synthesis. Although these are the biochemical changes that precede the attainment of hardiness, it is not certain that they cause hardiness, because other, nonhardy plants undergo the same changes when exposed to suboptimal temperatures (Chen & Li, 1982).

6.3   Freeze susceptibility and undercooling

Reference to Fig. 6.1 shows that organisms which are susceptible to freezing injury can survive only if ice formation anywhere in the tissues can be prevented. There are two possible means of achieving this: a colligative depression of the freezing point or the inhibition of ice nucleation (undercooling). In principle, the second alternative is to be preferred, because it does not require high concentrations of solutes which produce large increases in the viscosity of the aqueous substrate. As was discussed in Chapter 2, the temperature at which ice is nucleated in undercooled water depends on the volume of the aqueous phase and the presence of material which is able to catalyse the process. Thus, to lower the nucleation temperature, the organism would either have to redistribute its aqueous fluids into very small volumes, or it must contain substances which can inactivate the catalysts responsible for the nucleation of ice in the tissues.

Several of the above strategies are employed by various forms of life to prevent freezing in the tissue fluids. It must be borne in mind that for every method of protection there is also a lower limiting temperature at which the probability of nucleation rapidly increases and where freezing will occur; such freezing is invariably lethal. Evolutionary development has been such that a given organism achieves protection against freezing down to temperatures slightly below the normal minimum winter temperature to which it is exposed. Extremely low temperatures during the hardening process or abnormally cold winter conditions cause the death of such freeze susceptible species. The degree of protection in relation to the temperature of the natural environment is well illustrated by comparing polar fish species with overwintering insects and woody plants. Thus, the temperature of the Antarctic ocean is fairly constant at -1.5 °C which is approximately one degree below the freezing point of blood. The blood of Antarctic fish species freezes at -2 °C, equivalent to an undercooling of 1.5 deg. The body fluids of the beetle *Meracantha contracta* undercool to -10.3 °C when the larvae are in their fully hardened state (Table 6.2); this is sufficient to ensure their survival in their natural habitat. Some woody plant species native to temperate regions of North America and Asia exhibit the phenomenon of deep undercooling in which the water in the living xylem tissue is able to undercool to temperatures approaching -50 °C, substantially below the homogeneous nucleation temperature of ice in pure undercooled water. This ability to undercool is only observed during winter, when the tree is in its fully hardened state. During summer the tissue water will not undercool below -20 °C.

We shall now review several of the principles involved in freeze avoidance. Although some quite detailed physico-chemical and bio-chemical studies have been performed, we are still largely ignorant of the actual mechanisms at the molecular level whereby water can be prevented from freezing and made to exist for prolonged periods in a metastable state, and how these mechanisms are related to the seasonal hardening and dehardening of living tissues.

6.4 Antifreeze peptides

The first description of proteinaceous materials able to promote undercooling in tissue fluids dates from 1964. Since then such peptides and glycopeptides have been isolated from many fish and insect species.[†] The

---

† The term 'antifreeze protein' (AFP) was first used to describe the substances isolated from Antarctic fish species and has become common usage among fish physiologists and biochemists (DeVries, 1983). Unfortunately the same group of substances isolated from insects is now being increasingly described as 'thermal hysteresis proteins' (THP) by insect physiologists (Duman, 1982), presumably because they promote a freeze–melt hysteresis in the body fluids.

Confusion in the nomenclature is further compounded by the increasing use of the term 'freezing point' to describe the temperature at which *isolated* insect haemolymph is observed to freeze, while the temperature at which the water in the *intact* insect freezes is referred to as 'supercooling point.' It is of course the latter temperature which is of importance in freeze avoidance, and we shall refer to it as the freezing temperature. The thermal hysteresis is then the difference between this freezing temperature and the melting temperature.

first thorough study performed was of the glycopeptides found in the body fluids of Antarctic fish species in concentrations of 4% (w/v). The freezing–melting behaviour of aqueous solutions of various AFPs is shown in Fig. 6.2 (DeVries & Wohlschlag, 1969; DeVries, Komatsu & Feeney, 1970). The first AFP isolated was found to consist of a mixture of eight fractions, ranging in molecular mass from 2600 to 23 500 daltons, with the various fractions exhibiting different degrees of antifreeze activity. Chemically the fractions are closely similar: they consist of a basic tripeptide repeat unit ala–ala–thr, with the disaccharide side chain galactose-*N*-acetyl-galactosamine attached to each threonyl residue.

Later investigations on Arctic fish species led to the isolation of related AFPs, the main difference being that they did not possess the regular peptide sequences of the Antarctic AFP, nor did they contain carbohydrate residues. They did resemble the Antarctic AFP however in that two-thirds of the amino acid residues were alanine. More recently still, AFPs have been isolated from a whole range of overwintering insects and the amino acid composition determined. There are no obvious relationships between the amino acid compositions and the antifreeze activity. One distinguishing feature of the insect AFPs as a group is the much higher percentage of hydrophilic residues (40–60 mol %); (Duman, 1982).

The effectiveness of the AFPs in depressing the freezing point of water (and the body fluids) is truly remarkable. It is not due to an osmotic effect, since the melting temperature correctly reflects the very small depression

Fig. 6.2. *In vitro* undercooling potential of antifreeze proteins (AFP) isolated from different species. After Duman (1982) and references cited therein.

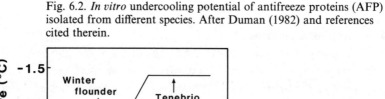
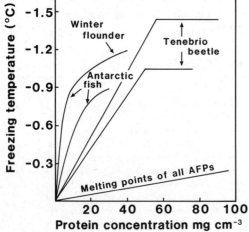

one might expect from a solute with a mean molecular mass of 10 000 daltons. The antifreeze effect is therefore due to an inhibition of ice nucleation in the undercooled body fluids. The AFPs also have a marked effect on the morphology of ice crystals grown in their solutions. Instead of the usual dendritic ice growth, needles are formed parallel to the crystal axis which is not normally the preferred axis of ice growth (DeVries, 1983). Despite considerable speculation about the mode of action of fish AFPs, all that is really known is that they can induce a degree of undercooling of up to 1.5 deg in dilute aqueous solutions. It is not even known whether they affect the *homogeneous* nucleation temperature of water (-40 °C).

The temperature of the polar oceans does not fluctuate much during the course of the seasons, which is probably the reason why the AFP levels in polar fish appear to remain fairly constant. The situation is of course different for the land based insects. The process of cold hardening presumably involves the seasonal synthesis and accumulation of AFP in the body fluids. Artificial cold hardening studies on the beetle *Tenebrio molitor* have shown that the degree of thermal hysteresis achieved depends both on the temperature of hardening and the photoperiod (Duman, 1982). Of the two factors the photoperiod seems to be the more important one at higher hardening temperatures, whereas at low hardening temperatures (5 °C) changes in the photoperiod do not significantly affect the freezing temperature. This is illustrated in Table 6.1. It is interesting that even under summer conditions (20 °C and long photoperiod) *Tenebrio* larvae maintain a significant ability to undercool. By contrast, *Meracantha* larvae lose this ability during the months of June to September (Duman, 1982).

6.5   Freezing point depressants

The degree of undercooling that can be achieved with the aid of antifreeze peptides is quite limited (see Fig. 6.2) and cannot by itself

Table 6.1. *Degree of undercooling and freezing temperature of* Tenebrio molitor *larvae under different hardening conditions*

Hardening conditions		Undercooling (ΔT °C)	Freezing temperature (°C)
Temperature (°C)	Light/Dark		
20	16/18	0.75	-7.7
20	6/18	1.67	-13.6
5	18/6	1.47	-14.0
5	6/18	1.35	-14.9

After Duman, 1982.

account for the very low freezing temperatures observed in overwintering insects. A clue is provided by the melting temperature which reflects the presence of water soluble solutes in the haemolymph. Relevant data for two species of beetles are provided in Table 6.2 (Duman, 1982). The haemolymph of *Uloma impressa* must contain substantial concentrations of solutes which accumulate during hardening and are metabolized during dehardening; these account for the melting point of -9.9 °C.[†]

Glycerol has been identified as a common melting point depressant, and the concentrations detected can account for the observed depression. That still leaves the very high degree of undercooling to be explained. Work with model aqueous solutions of various types has led to the suggestion of a linear relationship of a semi-empirical nature between the melting point and the nucleation temperature (Rasmussen & MacKenzie, 1972; Franks, 1981c). Thus, for a variety of low molecular mass solutes it has been found that $2\,\Delta T_m \sim \Delta T_h$, where T_h is the homogeneous nucleation temperature. For the few polymeric solutes that have been investigated, the proportionality constant lies between 2 and 4. The low undercooling temperature of *Uloma impressa* can be well accounted for on this basis. It is more difficult to explain the behaviour of *Meracantha contracta*, the haemolymph of which is not believed to contain polyhydroxy protectants.

The surprisingly low freezing temperatures reported for a variety of insects invite speculations about the mechanism responsible for inhibiting inadvertant nucleation at intermediate temperatures. One must conclude that catalysts that might promote heterogeneous nucleation are either

Table 6.2. *Melting and freezing temperatures of the haemolymph of* Meracantha contracta *and* Uloma impressa *in their hardened and tender states*

	M. contracta		U. impressa	
	Melting	Freezing	Melting	Freezing
Month	Temperature (°C)		Temperature (°C)	
February	-1.31	-10.3	-9.90	-21.5
June	-0.81	-3.8	-0.88	-6.2

After Duman, 1982.

    [†] It should be recalled that in an ideal solution the melting point of ice is
       depressed by 1.86 °C per mol of solute. Nonideality factors usually increase the
       magnitude of the depression for a given solute concentration, according to eqn.
       (3.11).

absent or have been rendered ineffective, perhaps by the high concentrations of polyhydroxy compounds. In this connection it is of interest to note that a period of starvation prior to exposure to subfreezing temperatures enhances the degree of undercooling that the insect can achieve (Sømme & Conradi-Larsen, 1977; Zachariassen, 1980). There is a clear correlation between low temperature mortality and feeding prior to cold exposure. The indications are that food particles provide catalytic sites for heterogeneous nucleation of ice in the gut. A more detailed discussion about the synthesis and metabolism of colligative freezing point depressants is deferred to the next chapter, because these substances are even more common in freeze tolerant organisms.[†]

6.6   Deep undercooling in plants

Many of the woody plant species that inhabit the temperate regions of Asia and North America have the ability to survive winter temperatures down to about $-40\ °C$. In such plants the hardening process renders the xylem resistant to freezing, a phenomenon referred to as *deep undercooling* (George, Becwar & Burke, 1982). In some plants, e.g. fruit trees (apple, pear, peach), some tissues are able to undercool, while others freeze. Where this occurs, e.g. in buds, a barrier must exist to the osmotic flux of water between the unfrozen cells and the extracellular fluid which contains ice and is therefore not in osmotic equilibrium with the cell water. The most useful techniques employed in the study of plant tissue freezing and undercooling are DTA (differential thermal analysis) or its more sophisticated development DSC (differential scanning calorimetry) and NMR (nuclear magnetic resonance) (Burke *et al.*, 1976). The two former methods monitor (directly or indirectly) the latent heat released during the crystallization of ice while the sample is subjected to cooling at a constant rate. NMR, on the other hand, provides a measure, albeit of a very indirect nature of the diffusional freedom of the water molecules in the tissue.[‡] A discontinuity in the measured parameter at some given temperature is taken as evidence of freezing of water. Both thermal and NMR methods

---

[†] Baust, Lee & Ring (1982) have published a comprehensive bibliography on insects at low temperatures, showing, incidentally, the rapidly growing interest in this subject.

[‡] Actually NMR utilizes the magnetic properties of nuclei; in particular the behaviour of ^1H, ^2H and ^{17}O which occur in water can be observed. After a disturbance of their magnetic states the nuclei return to equilibrium at rates determined by molecular interactions and the rates and types of molecular motion. In cases discussed here, the method is based on the very different rates of diffusion of water in the liquid and frozen states.

have been invaluable in monitoring the undercooling and freezing of water in plant tissues.

The thermal responses of water and plant tissues to cooling are shown diagramatically in Fig. 6.3 (Burke *et al.*, 1976). A bulk sample of water will, unless seeded with an ice crystal, undercool until ice nucleation is catalysed by some foreign particle present in the liquid. The degree of undercooling is not very reproducible but bears some relation to the volume. The important point to note is that it requires but a single nucleus to freeze the bulk sample, since the rate of crystal growth is very much higher than the rate of nucleation. The appearance of the freezing exotherm is as shown in (*a*), characterized by an abrupt onset of heat evolution. If the water sample is dispersed in an inert medium in the form of very small droplets (volume of the order of $< 100 \ \mu m^3$), the catalytic

Fig. 6.3. DTA freezing exotherms of (*a*) bulk water, (*b*) water dispersed in the form of a droplet emulsion in an inert oil, (*c*) tissue water in fully hardened hickory and (*d*) tissue water in hardened peach flower buds. After Burke *et al.* (1976).

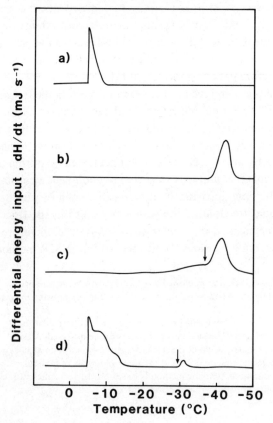

action of particulate impurities is suppressed, since the clean droplets vastly exceed those containing a catalysing particle. The sample then exhibits homogeneous nucleation, as shown in (*b*); for a homodisperse emulsion of water the shape of the exotherm is Gaussian.

Figure 6.3*c* represents the observed freezing behaviour of stem tissue from hardened hickory. The resemblance to (*b*) is striking and suggests that the water in this tissue freezes by homogeneous nucleation; hence the remarkably high degree of undercooling and the almost Gaussian shape of the exotherm. The actual freezing temperature (shown as −40 °C in Fig. 6.3*c*) is very sensitive to the degree of cold hardening; it is observed at −20 °C in samples of tender tissue, collected during summer. The arrow indicates the killing temperature which is seen to be closely associated with the onset of freezing.

The type of behaviour shown in (*c*) is rare; more commonly it is found that only a minor fraction of the water in a given tissue exhibits deep undercooling, whereas the bulk of the water freezes at a much higher temperature. This is shown for hardened peach flower buds in Fig. 6.3*d*. The tissue could also be classified as freeze tolerant, because it survives the freezing of the major portion of the tissue water. Deep undercooling affects only 5% of the water in the buds, but the freezing of this water is lethal and therefore sets the limit for cold hardening. The major exotherm (or series of three partly superimposed exotherms) represents the freezing of extracellular water (bark, cortex, phloem, cambium and non-living xylem cells), whereas the minor exotherm corresponds to the freezing of living cell water (xylem, ray and pith cells).

Most of the experimental studies of deep undercooling have been performed on vegetative parts of plants, but the phenomenon is also observed in reproductive structures, such as floral buds, cones and seeds. Exotherms similar to that shown in Fig. 6.3*d* have been observed in primordia of several species of *Cornus*, *Prunus* and *Rhododendron* (George, Becwar & Burke, 1982).

That the exotherms in Fig. 6.3*c* and *d* do indeed reflect undercooling, rather than colligative depressions of the equilibrium freezing point, has been established by determining the melting points of the tissue water. In every case the ice melts in the neighbourhood of −2 °C, corresponding to an osmotic solute concentration of approximately 1 osm kg^{-1}. NMR studies on undercooled tissue have confirmed these conclusions by indicating that the water has a much higher degree of diffusional freedom than would be expected for water in a concentrated solution. Undercooled tissue that was kept at some intermediate temperature was found to freeze after long periods, in agreement with the theory of nucleation. It has also been observed, but is probably not of any ecological consequence, that by maintaining hardened tissue at subzero temperatures for long periods, deep undercooling can be further enhanced, and in some cases the tissue can be rendered completely freeze resistant, i.e. the exotherm identified with death disappears.

Although the experiments here described illustrate the variety of responses to cold stress, they provide no information about the mechanism of hardening or the detailed means by which small domains of water can avoid freezing even in the presence of ice elsewhere in the tissue. What is certain is that undercooling is not of a colligative origin, caused by high concentrations of protective solutes. The role of the so-called unfreezable water cannot be ruled out, as cooling experiments on seeds have shown (Juntilla & Stushnoff, 1977). Thus, lettuce seeds containing 40% water undercool to −15 °C before freezing. A reduction of the water content to 25% increases the undercooling to −25 °C and seeds with a water content reduced to 16% exhibit no freezing at all. Similar observations have been made with rice and wheat seeds. Deep undercooling of seeds in their natural setting has also been reported: the seed of dwarf mistletoe is able to undercool to −30 °C but is killed once freezing does occur.

Finally, little is known about the anatomy and histology of undercooling of tissues in the presence of ice. Questions have been raised about the significance of laboratory experiments which are almost invariably performed under conditions of 'rapid' cooling (compared to field conditions) such as might well promote undercooling, the effects being, however, quite unrelated to resistance mechanisms under ecological conditions.

6.7   Other freeze avoidance mechanisms

Organisms whose normal habitats are the polar oceans require permanent protection against freezing because the temperatures rarely rise above the equilibrium freezing points of body fluids. Various methods have been reported, some relying on the synthesis of protecting substances and others on the reversible dissociation of large macromolecules normally present in the haemolymph.

Molluscs which live at the water line of icebergs are often trapped by the advancing ice in pockets of undercooled water. They respond by secreting mucopolysaccharides which prevent the water globule from becoming increasingly freeze concentrated in salt as the temperature drops.

Freeze avoidance by protein dissociation has been observed in the gastropod *Helix pomatia* (Douzou & Balny, 1974), the respiratory system of which contains copper haemocyanins of exceptionally high molecular masses, approaching 9×10^6 daltons, composed of twenty subunits. These structures are sensitive to changes in pH and the partial pressure of oxygen and they can dissociate into the individual subunits which appear to be identical, with molecular masses of 4.5×10^5 daltons. *In vitro* experiments have shown that a 75% dissociation produces an undercooling of the haemolymph to -10 °C. Here again the effect cannot be of an osmotic origin, because the *melting* point of the ice is unaffected. The protein dissociation is accompanied by a marked increase in the viscosity of the haemolymph, but it is not clear how this alone could produce such a marked undercooling. Among the possible mechanisms are an action akin to that of the antifreeze peptides, although globular proteins are not thought to provide antifreeze activity, or the very large increase in protein–water interface which accompanies the dissociation. This might increase the proportion of 'unfreezable' water. However, the fact that the water does eventually freeze makes this an improbable alternative. Another direct consequence of the creation of a large protein–water interface is the change in the physical nature of the system: possibly a disperse structure is created in which the water is held in the interstices between the protein subunits. Such an effect would reduce the nucleation rate of ice in the undercooled haemolymph. Similar protein dissociation effects have been observed in crustaceans. The partial dehydration of the cell fluids during freezing triggers the dissociation which then prevents further dehydration

Fig. 6.4. Expanded cortex model of a resistant bacterial endospore. Hardening is accompanied by the synthesis of the negatively charged cortex which dehydrates the core thus rendering it heat (and freeze?) resistant. Reproduced from Gould & Measures (1977).

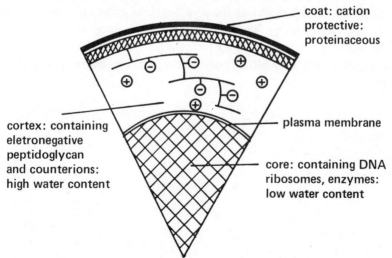

and promotes undercooling. Perhaps such survival mechanisms are more common than has been believed: it is generally found that only 60–70% of the body water of intertidal animals freezes. On the other hand these animals do not synthesize glycerol or other polyhydroxy compounds as freeze protectants.

Finally, a particularly refined mechanism for inducing freeze avoidance is that employed by spore-forming microorganisms. When the cell is put under stress, e.g. by starvation, it rapidly responds by synthesizing a negatively charged polymer (peptidoglycan). This macromolecular structure is highly chain branched and cross linked. Its high charge density causes it to expand and to 'squeeze' the water out of the centre of the cell into the polymer structure where the water is held by the COO^- groups, as shown diagrammatically in Fig. 6.4 (Gould & Measures, 1977). The core of the cell which contains the organelles is thus partially dehydrated which renders it unfreezable. Simultaneously with the build-up of the peptidoglycan cortex the cell also produces appreciable concentrations of trehalose, a glucose based disaccharide which is known to protect proteins against denaturation and which also inhibits the phospholipid bilayer phase from undergoing the transition to a hexagonal structure upon dehydration (Crowe & Crowe, 1981). The cell is now protected against the injurious effects of heat and freezing and exhibits the ultimate in biological resistance. However, while in the dormant state, it stays alert to the trigger which can cause germination and the eventual recovery of full metabolic activity. Actually spore formation is accompanied by several additional biochemical changes, but we emphasize those that result in the redistribution of water (without an overall dehydration) from those domains where freezing would be harmful, to the peptidoglycan structure where freezing does not cause injury. The remarkable effectiveness of this form of protection is demonstrated by the fact that the DNA and enzymes isolated from spores show none of the resistance which is so characteristic of the intact cell.

7

Freeze tolerance in living organisms

7.1　Freeze tolerance: a misnomer?

In the physiological and ecological literature a distinction is drawn between those forms of life which can survive freezing and those which cannot. The latter, in order to survive, must possess the means for seasonal protection where environmental temperatures fluctuate about the freezing point of water. Freeze susceptible animals which inhabit those regions where the temperature is always below the equilibrium freezing point of their body fluids require permanent protection. The term freeze tolerance is commonly applied to organisms which can tolerate ice in their tissues. In a sense this is misleading, because such tolerance never extends to ice growth within the living cells. The hardening processes which give rise to freeze tolerance must therefore be such as to render cytoplasmic water unfreezable. As far as the living cell is concerned, freeze tolerance does not exist: survival at the cellular level is conditional upon freeze avoidance, however achieved. This must be borne in mind in discussions of the various processes involved in rendering living organisms tolerant to ice crystallization in their tissues.†

Biophysical and biochemical investigations into freeze tolerance and frost hardening fall into three distinct categories: those in which emphasis is placed on membrane related phenomena, those which describe metabolic modifications during the hardening period and those which concentrate on changes in the water relationships within the organism. That hardening is accompanied by changes in the membrane lipid composition is now well established. The synthesis of various low molecular weight solutes, usually

---

† From time to time isolated reports appear which describe nonlethal intracellular freezing. However, such observations always refer to cells cooled under carefully controlled conditions in the laboratory, rather than to whole organisms or tissues subjected to climatic fluctuations.

by the hydrolysis of starch, is also a universal feature of the hardening process. The search for a single stress/strain event as a critical cause of freeze injury is hardly rewarding, yet claims of this type are in the habit of being made. For instance, the view has been expressed that freeze tolerance is entirely due to the colligative action of the various low molecular mass carbohydrates on the freezing temperature of tissue fluids (Santarius, 1982). The evidence against such simplistic interpretations is overwhelming: detailed comparisons on hardened and tender plants have indicated that identical metabolic changes occur during simulated hardening conditions and that the nature of the injuries are also identical, with the difference that in the hardy varieties the injury symptoms become apparent at lower temperatures than in the tender varieties (Chen & Li, 1982).

7.2   Phenomenology of frost hardening and injury

The changes that take place during freezing involve a combination of concentration effects with those other effects which are due purely to a reduction in the temperature. Injuries due to freezing are therefore more difficult to analyse than injuries caused by chill or undercooling in the absence of freeze concentration and the resulting osmotic stresses across membranes. As was discussed in Chapter 5, carefully controlled experiments on model systems, such as phospholipid vesicles, and on single cells (usually human erythrocytes) have highlighted the sensitivity of the plasma membrane to freezing stresses. In this section we review the injury symptoms and the conceptual models advanced to describe, if not explain, frost hardening.

There is now a general consensus that membrane mediated processes are very sensitive to freeze damage. Early damage is observed at the mitochondrial and chloroplast membranes, followed by the inability to fix CO_2 and finally by the complete inactivation of the thylakoid membrane but not, apparently, through the accumulation of ions during freeze concentration (Gusta, Burke & Kapoor, 1975; Heber *et al.*, 1981). Much of the early damage is reversible: NMR studies on winter wheat have shown that all the water that will freeze, does so above $-10\,°C$, so that further cooling is then not accompanied by any more freeze concentration. Yet the plant is killed only at temperatures approaching $-30\,°C$, as monitored by the massive leakage of ions. A similar lack of correlation between freeze dehydration and lethal temperatures has been observed for other winter hardy plants. While freeze dehydration proceeds gradually with decreasing temperature, loss of recovery occurs abruptly, usually over a very narrow temperature range.

Changes that occur in the thylakoid membrane lipids of plants during the period of frost hardening have been subjected to intensive studies. The comprehensive investigations by Heber and his colleagues on spinach chloroplasts deserve special emphasis (Heber *et al.*, 1981). The results can be summarized as follows:

(1) Natural hardening and simulated hardening in the greenhouse produce the same changes in the lipid profile, whereas control specimens, maintained at a constant temperature, do not exhibit any changes in the fatty acid chain distribution or the total lipid concentration;

(2) During the hardening period (September to January) a marked increase occurs in the total lipid content;

(3) The ratio of mono-:diunsaturated fatty acids decreases from 2.6 to 1.6;

(4) There is a general conversion of phospholipid to glycolipid, reaching 60–90% in the fully hardened state;

(5) The overall degree of unsaturation also increases, up to 90% of the total fatty acids;

(6) All the above changes take place gradually during the hardening period and are not reversed until dehardening begins during spring.
The lipid profile is unaffected by short term fluctuations in the environmental temperature. The above changes have been observed in a wide variety of tissues (roots, stem, buds, leaves) and many different plant species.

The popular interpretation of the observed membrane lipid changes is in terms of a requirement of membrane flexibility (fluidity) for survival; the merits of this hypothesis have already been discussed. However, it is not at all clear how the membrane changes help to protect an organism against frost damage and produce resistance. Experiments on frost sensitive membrane systems (thylakoids) taken from leaves of different degrees of frost hardening have failed to show a direct relationship between an enhanced lipid content and an increase in frost tolerance. Rather, a correlation was observed between lipid increases and sudden temperature changes.

By monitoring the degree of resistance of cabbage chloroplasts (i.e. the lowest temperature at which photosynthesis could take place) as a function of the minimum ground temperature, Critchley (1976) was able to show that over a period of six months the degree of resistance followed the changes in the ground temperatures, with the killing temperature of the thylakoids always lying about 10 °C below the minimum ground temperature recorded a few nights previously. The results are shown in Fig. 7.1. The surprising feature is the rapid adaptation of the thylakoids to short term fluctuations in temperature, e.g. during January when a relatively mild period was followed by a severe cold. Also shown in Fig. 7.1 is the change in total lipid content (relative to the chlorophyll content). It is seen that the rise in the lipid content does not commence until November, when the mean ground temperature has already fallen by 10 °C, and that the dramatic increase in total lipids between November and January is not correlated with the changes in ground temperature during the same period; nor is it correlated with the development of frost resistance, because the leaves have already developed a substantial degree of resistance before changes are observed in the lipid content and composition.

Turning once again to the calorimetric monitoring of freezing, reference to Fig. 6.3*d* shows a typical example of what is usually described as freeze tolerance. The significance of the three major exotherms is unknown but the observations are indicative of three independent crystallization processes in three different regions of the flower bud. From the relative areas of the various exotherms it is deduced that up to 95% of the tissue water freezes at high subzero temperatures, with the positions and shapes of the three major exotherms not substantially affected by the hardening process. The minor exotherm, however, which is observed at −15 °C in tender apple xylem, shifts to −20 °C during the early autumn and to −30 °C after a brief exposure of the tissue to 0 °C; it reaches its lowest temperature after

Fig. 7.1. Minimum ground temperatures and frost resistance of cabbage leaves during the period of hardening and resistance (September–March). Also shown is the total lipid content of the thylakoids. After Critchley (1976).

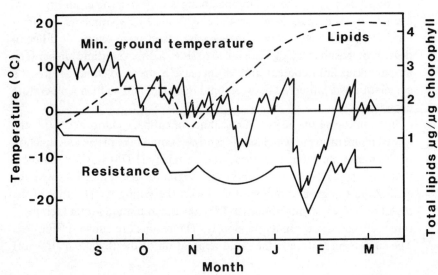

exposure of the tissue to $-5\ ^\circ C$. A temporary increase in the temperature results in partial dehardening of the tissue, but its ability to recover full resistance is not lost until the buds begin to swell in spring.

The fact that the temperature of the minor freezing event changes with degree of hardening but that the area of the exotherm is constant throughout the year indicates that, whatever intracellular freeze concentration may occur during the freezing of the extracellular water, it reaches an upper limit, so that eventually the hardened living cell becomes isolated from its surroundings. Its freezing cannot be seeded by extracellular ice and it is not subject to an osmotic stress which might cause its complete desiccation.

Apparently genuine freeze tolerance is observed in woody plants which range into the Arctic regions where temperatures below $-40\ ^\circ C$ are not uncommon. The DTA thermograms exhibit only the three high temperature exotherms. On further cooling no ice crystallization is observed, the tissue being resistant down to the lowest temperatures that have been studied ($-196\ ^\circ C$) (George, Burke, Pellett & Johnson, 1974). It appears that the dehydration of the living cells proceeds to such an extent that no freezable

Fig. 7.2. Representative model for the complete plant cycle: growth, maturation, hardening, dormancy, dehardening; for details see text. The broken curve represents the development and maintenance of frost resistance, also shown as a hardiness index in the lower portion of the figure. After Fuchigami *et al.* (1982).

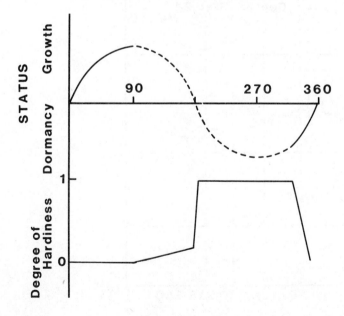

water remains. Strictly speaking, therefore, the process is really again one of freeze avoidance (by the living cells), the freeze tolerance being confined to those parts of the plant where freezing does not cause damage. Although the mechanistic details of frost hardiness are poorly understood, the complex interplay between growth and hardening has been observed in many different species so that it is possible to model the various features of the cyclic processes. Thus, Fuchigami *et al.* (1982) have developed the model shown in Fig. 7.2 on which the various events in the plant's development can be identified. The first quadrant of the cycle corresponds to the rapid growth stage, affected mainly by temperature and level of illumination. During this period the plant is susceptible to frost. Maturity begins at 90°, when the plant becomes susceptible to suboptimal environmental factors, e.g. short day length. During this period the hardening process begins, and plants defoliated after the point corresponding to 180° can still develop hardiness. The winter quiescence stage is characterized by no growth but maximum resistance. Dehardening begins at 270°, but until the buds begin to break in spring (0°), dehardening can still be reversed. The various points on the cycle can be determined experimentally;

Fig. 7.3. Correlation of measured daylength and temperature with the prediction of the cyclic model (Fig. 7.2), as applied to red-osier dogwood. After Fuchigami *et al.* (1982).

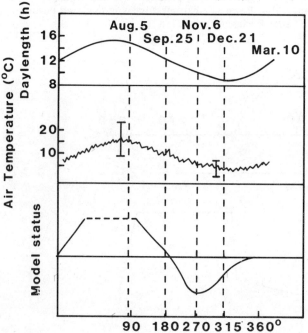

it is found that the times taken by different plants to reach a given point on the cycle differ considerably. For instance, the 180° point can be established by measuring stem growth after defoliation which, starting from a low value at 90°, reaches a maximum at 180°. Conversely, ethylene production decreases from a high value at 90° to a low, but stationary, value at 180°. The 270° point is estimated by exposing the plant to higher temperatures and longer photoperiods for increasing periods until dormant buds begin to grow, or by the determination of the gibberellic acid concentration required to achieve regrowth.

The success of the growth/dormancy cycle in accounting for the actual behaviour of a plant is shown in Fig. 7.3 where daylength and temperature changes are compared to growth and dormancy changes determined for red-osier dogwood, grown in Oregon. The model is well able to account for the observed developmental cycle.

7.3   Photosynthesis at subfreezing temperatures

Despite extensive investigations into the effects of subfreezing temperatures on photosynthesis in plants, the mechanisms of injury and hardening are obscure (Levitt, J., 1980). For temperatures above 0 °C it is found that photosynthesis will adapt to suboptimal conditions. The period required for hardening is usually of the order of days, or even weeks, but some plants (e.g. wheat) can harden rapidly, even within 24 hours. The optimum temperature for photosynthesis usually correlates with the mean maximum temperature of some preceding period (10 days in the case of *Eucalyptus pauciflora*) (Slayter & Morrow, 1977); see also Fig. 7.1.

All plants, unless subjected to a hardening period, are killed by exposure to temperatures below -2 to -6 °C. That is to say, photosynthesis is completely inhibited under these conditions. If, on the other hand, the plant is maintained under frost hardening conditions (low temperature and short photoperiod), then the potential rate of light saturated photosynthesis is observed to decrease. However, it is not clear whether such a reduction is simply the result of low temperature on the chemical reaction rates or whether it corresponds to the onset of dormancy. This type of response to low temperatures is characteristic of most plants that are able to survive freezing temperatures (e.g. wheat, barley, rape), but it is particularly well established in conifers, where photosynthesis stops at the temperature at which the needles freeze (approximately -5 °C). If the first frost of the season is followed by a period of warmer weather, photosynthesis recovers after a few days. If, on the other hand, frost conditions persist for an extended period, then photosynthesis stops throughout the winter (Öquist, 1983).

At first sight it seems likely that the loss and recovery of photosynthesis might be due to stomatal closure and opening, because they are affected in a similar manner by temperature. However, there are no clear examples of a temperature induced stomatal closure being a primary cause of photosynthetic inhibition. Also, it is found that transpiration recovers before CO_2 uptake recommences after a brief period of frost.

Assuming, therefore, that an exposure to low temperatures affects the actual mechanism of photosynthesis, the problem becomes one of identifying the particular reaction (or reactions) where such changes occur. It has not yet been possible to construct a credible model because of the complexity of the various interactions involved. So far all that has been achieved is to distinguish between reversible and irreversible processes, but there are still uncertainties as to which of the biochemical effects observed are direct consequences of the low temperature and which might only be related indirectly to photosynthesis.

Purely photochemical processes are unaffected by temperature. On the other hand, all enzyme catalysed reactions are sensitive to temperature and may be especially sensitive to freezing (see Chapter 4) because of the concentration effects involved. Indeed, low temperature studies on spinach chloroplasts have shown that all enzyme catalysed reactions stop at the freezing temperature but that choloroplasts, undercooled to $-15\,^{\circ}C$ in ethane diol/water mixtures, are able to function (Cox, 1975). Although freezing leads to an immediate inhibition of photosynthesis, this is not the case for the partial photosystem reactions. It therefore appears that the site of inhibition is located between the photosystems. The freeze inhibition of photosynthesis closely resembles the effect of herbicides which block photosynthetic electron transport between quinone and plastoquinone in photosystem II. Other symptoms of freeze damage include the partial destruction of chlorophyll–protein complexes and a change in the chlorophyll a/b ratio (Öquist & Martin, 1980).

Detailed studies on a variety of conifers lead to the conclusion that the inhibition of photosynthesis is controlled by the electron transport properties of thylakoids at the plastoquinone site. Subfreezing temperatures (above the killing temperature) do not by themselves inhibit photosynthesis, although they do reduce the capacity for photosynthesis. The same conditions, but in the presence of light, do inhibit photosynthesis, possibly through photoinhibition and photooxidation of membrane proteins. The low temperature therefore sensitizes the photosynthetic apparatus to photooxidation by light. It must be emphasized that these stress symptoms, although severe, do not kill the plant and the damage can be repaired.

Öquist (1983), proceeding from a careful analysis of the available experimental evidence, identifies the reasons for lack of agreement and lack of progress in this area as follows: (1) Unwarranted comparisons are made between species which exhibit different degrees of adaptation. (2) Too much reliance is placed on experiments on isolated organelles or isolated enzyme systems. In the first place the experimental isolation procedures may give rise to the release or synthesis of inhibitors, but even where this is not the case, *in vitro* results may be misleading. For instance, experiments on isolated preparations may indicate that photophosphorylation is inhibited by low temperatures, whereas in the intact plant inhibition is found to be due to electron transport. (3) There is as yet no agreement as to which of the observed effects can be directly related to the influence of temperature on photosynthesis. (4) Too many conclusions are drawn from limited studies of partial reactions. (5) Not enough attention is paid to standard conditions in hardening experiments.

7.4   Biochemistry of freeze protection

Mention has already been made of the synthesis of freezing point depressants as a means of inducing frost hardiness. The substances most frequently encountered are sugars and sugar alcohols, derived from starch. Amino acids and various intermediary metabolites have also been implicated in the promotion of freeze tolerance. Increases in the concentrations of low molecular weight solutes not only lower the freezing point of the tissue fluids but also reduce the total amount of water which is able to freeze, and thus the freeze concentration factor (see Chapter 3).

The seasonal changes in the solute and metabolite profile of the freeze tolerant gall fly larva *Eurosta solidaginis* have been recorded and are shown in Fig. 7.4 and Table 7.1 (Storey, Baust & Storey, 1981). The most striking features are the increases in the levels of sorbitol and glucose (by more than two orders of magnitude), although several of the other glycogen metabolites and amino acids also show significant concentration increases. In the fully hardened state the larvae are able to survive at temperatures down to $-40\ °C$, although extracellular freezing is observed at $-8\ °C$ (this is the nucleation temperature of ice in the haemolymph and *not* the equilibrium freezing point). Other insects show a similar relationship between build-up of polyhydroxy compounds and degree of cold hardening.

The time resolution of the various biochemical processes during the period of hardening indicates that glycerol production begins during August/September, long before the insect experiences the first cold exposure. The glycerol level reaches its plateau value of 650 mM in the haemolymph before the synthesis of sorbitol begins. Glucose and sorbitol are only produced after the first cold exposure and they reach their midwinter levels when the temperature approaches the extracellular nucleation point (−8 °C). Minor increases occur over the same period in the concentrations of other starch metabolites and of the total amino acids, especially proline and alanine. However, unlike the sugars and sugar

Table 7.1. *Changes in metabolic levels in* Eurosta solidaginis *larvae during cold hardening*

Concentrations in μmol/g wet weight

	15 °C	0 °C	−5 °C	−30 °C
Glycogen	381	258	166	113
Glycerides	167	170	169	159
Protein	81		unchanged	
Glycerol	154	232	237	238
Sorbitol	1	42	97	147
Glucose	0	16	34	29
Trehalose	55	64	63	71
Proline	32	42	57	56
Glycerol-3-phosphate	0.2	0.9	1.6	1.0
ATP	2.2	2.2	2.2	1.6
Lactic acid	0.2	0.1	0.1	0.5

After Storey, Baust & Storey, 1981.

Fig. 7.4. The cold-induced conversion of glycogen into sugars and sugar alcohols by *E. solidaginis* larvae. After Storey *et al.* (1981).

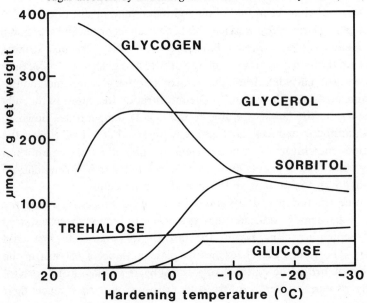

alcohols, the excess amino acids are produced by *de novo* synthesis and not by protein hydrolysis. The transient increase in the glutamate level may be due to its role as an intermediate in the synthesis of proline.

Other chemical changes during hardening include a decrease in the ATP level and a significant increase in lactic acid, indicating some anaerobic carbohydrate catabolism. The various hardening processes therefore appear to induce a state of dormancy, possibly resulting from a restricted supply of oxygen in the frozen state. Figure 7.5 shows the marked difference in the oxygen uptake in the frozen and undercooled state of a silk moth pupa (*Hyalophora cecropia*) during diapause. The discontinuity at the extracellular nucleation temperature reflects the onset of freeze concentration in the haemolymph which in the unprotected insect would be lethal. The high concentrations of polyols and sugars render the various biological structures resistant to a substantial degree of dehydration. On the other hand, the process of freeze concentration is accompanied by a marked increase in the viscosity of the residual liquid phase which leads to reductions in the rates of biochemical reactions.

The biochemical details of cold hardening have been analysed by Storey (1983). They reveal a complex interplay between the environmental conditions and the various enzyme catalysed processes, eventually resulting in the acquisition of cold resistance by the larva. The obvious routes for the

Fig. 7.5. Oxygen uptake by diapausing silk moth pupae in the undercooled and frozen states, showing the injurious effects of freezing.

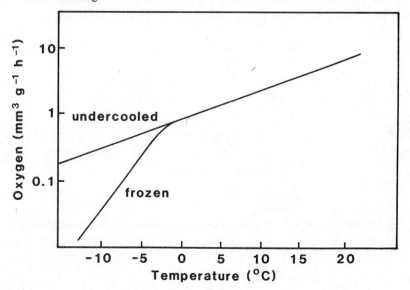

synthesis of glycerol and sorbitol are via the glycolytic pathway at the triose phosphate and hexose monophosphate levels respectively. Glycerol, once it has been produced during the early autumn, cannot be metabolized and remains in the haemolymph all through the winter, possibly because of the absence of a glycerol kinase enzyme. Sorbitol, by contrast, is a variable protectant: during transient warm periods it is reconverted to glycogen. The glycolytic enzymes which are involved in the production of cold protectants are stimulated by low temperatures, but no alteration in the isozymic forms of the glycolytic enzymes can be observed as a result of exposure to hardening temperatures. As might be expected, glycogen breakdown rates are closely correlated with the appearance of the low molecular mass protectant substances. Thus, glycogen hydrolysis is fast over the temperature range 15–10 °C, while glycerol is produced. The hydrolytic rate drops as the temperature is lowered from 10 to 5 °C, but increases dramatically with a further drop in temperature to −5 °C, corresponding to the production of sorbitol.

The scheme shown in Fig. 7.6 represents possible pathways for the production of glycerol and sorbitol. For the synthesis of glycerol the block would presumably be at or below the glyceraldehyde-3-phosphate dehydrogenase level, whereas for sorbitol production phosphofructokinase (PFK) acts as regulator. Since glycerol production is complete by the time of the first cold exposure, cold active PFK is not required. Other insects, e.g. *H. cecropia*, which only produce glycerol as a result of cold exposure, require PFK which can function at subzero temperatures.

Details of the cold inhibition of PFK are well established (see Chapter 4). The inhibition is particularly marked in the case of *E. solidaginis*, as

Fig. 7.6. Possible pathways of polyol synthesis in *E. solidaginis* larvae; for details see text. After Storey *et al.* (1981).

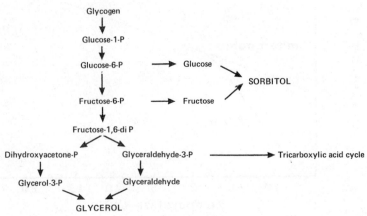

witness the remarkably large activation energy, 83 kJ mol^{-1} and $Q_{10} = 3.6$, compared to the more usual value of 50 kJ mol^{-1} and $Q_{10} = 2$. In addition to the extreme temperature sensitivity, the enzyme exhibits a reduced affinity for its substrate, fructose-6-phosphate, in the presence of sorbitol and glycerol-3-phosphate, both of which compounds are accumulated in the larva at low temperature (see Fig. 7.4). PFK thus affords a good example of the economy of function by which the larvae exploit temperature and modulator effects in order to use their metabolic make-up for two different functions at different temperatures.

7.5   Physical chemistry of freeze protection

The role played by polyhydroxy compounds in the stabilization of native proteins against dehydration (concentration) and extreme temperatures has already been discussed in Chapter 4. Another vital aspect of freeze accommodation may well involve the response of the plasma membrane to the stress caused by osmotic dehydration of the cell by extracellular freezing. Quite apart from thermotropic lipid phase transitions which might alter the permeability properties of membranes, an obvious consequence of loss of cell water must be the reduction of the cell volume. Given that the total amount of lipid remains constant, such a volume reduction must produce either a closer packing of the lipids in the bilayer or a drastic change in the shape of the cell. The former effect would be expected to give rise to lipid–lipid repulsions in the plane of the membrane, whereas the latter effect must produce domains within the membrane where the radius of curvature is appreciably larger than the value associated with the cell in osmotic equilibrium with its surroundings.

A combination of classical surface chemical techniques and cryomicroscopy on protoplasts has been particularly successful in the elucidation of the stresses imposed on membranes during freezing and the manner in which membranes are rendered resistant to freezing damage by cold hardening processes. A widely held view is that freeze injury is the direct consequence of cell shrinkage below a minimum critical volume (the minimum volume hypothesis), rather than of the high intracellular solute concentration (Meryman, 1974). The response of the membrane to cell contraction has been studied by extracting the lipids, spreading them as a monomolecular layer on an aqueous substrate and estimating the compressibility of such monolayers (Williams, Willemot & Hope, 1981). A typical surface pressure/(area)$^{-1}$ plot is shown in Fig. 7.7 for a pure lipid. During the initial compression the pressure (π) varies in a linear manner with the inverse area (A) occupied by the lipid molecules, corresponding to a two-dimensional analogue of Boyle's law for an ideal gas. When a certain critical pressure is reached, the lipid film becomes incapable of further compression. The plateau region in the π/A^{-1} diagram represents the collapse of the lipid film. On decompression the linear relationship is once again observed, but there is now less material in the lipid film. However, the extrapolated value of π, corresponding to $A^{-1} = 0$ (a two-dimensional ideal gas), is identical to the initial value, showing that the physical nature of the lipid at the surface has not changed.

By extracting lipids from hardened and unhardened wheat and determining the π/A^{-1} characteristics, Williams *et al.* (1981) established a general behaviour resembling that shown in Fig. 7.7. However, the behaviour of lipids isolated from hardened cells differs from that of unhardened cells in the area of the compression/decompression loop, in the sense that hardening results in a reduction of the film collapse phenomemon. In other words, lipids isolated from hardened cells are better able to resist compression.

This resistance is particularly marked with a very hardy cultivar, Kharkov winter wheat; the π/A^{-1} curves are shown in Fig. 7.8. The pronounced curvature in the initial compression plot corresponds to a loss of lipid from the surface, and the limiting pressure for the hardened wheat

Fig. 7.7. Idealized surface pressure (π) (area)$^{-1}$ profile for a lipid monolayer on an aqueous substrate. At a compression of π_0 the film becomes unstable and further compression leads to a removal of lipid into the substrate. During decompression the lipid is reincorporated into the film in an unchanged state. The complete reversibility of the compression/expansion cycle is indicated by the common extrapolated intercept π_∞. After Williams *et al.* (1981).

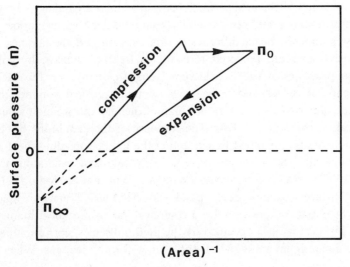

lipids is considerably lower than the value of 25 mN m^{-1}, characteristic of unhardened wheat lipids. The other remarkable feature of the hardened Kharkov wheat lipids is the closed compression/decompression loop, indicating that the material which was removed from the surface film during compression is reincorporated during decompression. The effects associated with lipid removal and reincorporation become even more pronounced when extracted cell fluids are substituted for water as the aqueous substrate. On the other hand, the effects are largely lost when purified lipids are used on a water substrate. It seems, therefore, that the hardened cells contain substances which facilitate the reversible removal of lipids during the stress imposed by cell shrinkage. The ability to replace the stored lipid into the membrane during deplasmolysis would account for the high degree of freeze resistance exhibited by Kharkov wheat.

Although the surface compression measurements do not in themselves provide any information about the mechanism whereby lipid can be removed from, and restored to, the membrane in response to changes in the cell volume, model experiments on different types of aqueous substrates suggest possible means by which this could be achieved. Thus, in the presence of the disaccharide trehalose, the spread phospholipid film takes

Fig. 7.8. Experimental π/A^{-1} profiles of the total extracted lipids from Kharkov winter wheat on a water substrate: A – unhardened; B – after 28 days hardening. Curve C corresponds to B, but the aqueous substrate is composed of the cell extract: the flattening in the curve indicates that the reversible loss of lipid from the surface has been facilitated. After Williams *et al.* (1981).

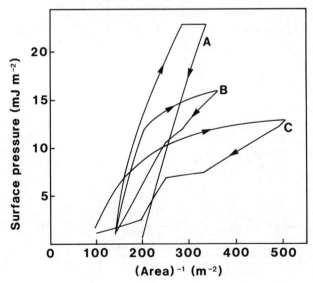

on a more expanded appearance. In other words, the addition of trehalose produces an effect that mimics the natural cell fluid (see Fig. 7.8). It is also known, however, that trehalose is able to inhibit the phase transition from a bilayer type structure to a hexagonal structure which phospholipids suffer as a result of dehydration to water contents of < 0.25 g water/g phospholipid (Crowe & Crowe, 1981). Trehalose has also been identified as a natural protectant against severe dehydration in a variety of organisms which may be exposed to periods of extreme water stress (e.g. bacterial spores). In the light of all the above observations, it is attractive to argue that trehalose, through its ability to preserve the 'natural' phospholipid bilayer structure, is able to affect the phospholipid removed from the membrane during cell volume reduction so that it remains in a form (the bilayer rather than the hexagonal phase) that can readily be reincorporated into the membrane during deplasmolysis.

The lipid monolayer studies provide indirect evidence that cold hardening involves some process that enables lipids to be transferred from the

Fig. 7.9. The effects of cooling rate and temperature on the growth of intracellular ice in tender (*a*) and hardy (*b*) rye protoplasts. Reproduced from Steponkus *et al.* (1983).

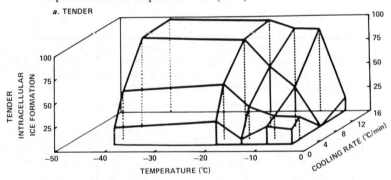

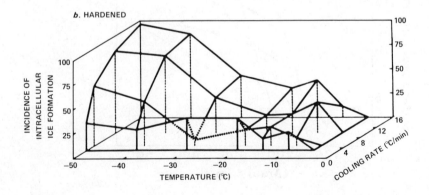

membrane to the cytoplasm in a manner such that during thawing and deplasmolysis the lipids can be rapidly reincorporated in the membrane, thus preventing the irreversible leakage of cell material. It is suggested that the driving force for such lipid transfer is the lateral pressure built up in the membrane during cell shrinkage.

Rather more direct, *in situ* observations on cell membranes during freezing have confirmed the transfer of membrane lipid during freezing. Thus, Steponkus and his colleagues have studied isolated plant protoplasts with the aid of a cryomicroscope, the stage temperature of which is programmed to given cooling and warming rates. Their work on tender and hardened Puma rye (*Secale cercale* L. cv. Puma) has laid the foundation for an understanding of the response of the plasma membrane to subzero temperatures (Steponkus, Dowgert & Gordon-Kamm, 1983). Figure 7.9 shows the complex relationships between temperature, cooling rate and the incidence of intracellular freezing. Taking a standard cooling rate of 16 deg min^{-1} and a standard temperature of -40 °C, it is seen that 100% of tender cells suffer intracellular freezing, whereas only 80% of the hardened cells freeze; it requires -50 °C to kill all the hardened cells. It is also found that tender cells are liable to ice seeding (by extracellular ice) at -15 °C, whereas hardened cells are able to resist seeding down to a mean temperature of -42 °C, suggesting that the membranes of hardened protoplasts provide effective barriers.[†]

The involvement of hypertonic stress in injury to membranes is demonstrated by monitoring the effects of increased osmolality of the suspending medium at normal temperatures. Here again it is found that hardened cells are better able to resist hyperosmotic conditions than are tender cells, but this may in part be due to the endogenous synthesis of protectant solutes during hardening. A significant difference in the response of tender and hardened protoplasts to hyperosmotic conditions appears to be in the manner in which lipid is removed and stored: tender cells secrete excess lipid by endocytosis and cannot reassimilate such material during deplasmolysis, whereas hardened cells secrete lipids by exocytosis and are

[†] Here, as elsewhere in the literature devoted to cell freezing, confusion surrounds the use of the terms *seeding* and *nucleation*. It must be understood that nucleation is an intrinsic process which relies on random density fluctuations in the liquid, but in the absence of the crystalline phase. Seeding, on the other hand, is produced by a crystal which is introduced into the undercooled liquid phase. In freezing experiments on cells, where the extracellular liquid is partly frozen, the likelihood of seeding is much greater than the likelihood of spontaneous nucleation of ice within the cytoplasm. If *nucleation* of ice in undercooled cells is to be studied, this had best be done in the *total* absence of ice in the medium surrounding the cells (Franks *et al.*, 1983; Mathias, Franks & Trafford, 1984).

Fig. 7.10. Electron micrographs of isolated rye protoplasts subjected to various treatments. *a* and *b*: Scanning electron micrographs of unhardened (in hypertonic 1.00 osm sorbitol) and hardened (in hypertonic 2.53 osm sorbitol) protoplasts. Extruded lipids are denoted by E and by arrows. *c* and *d* are transmission electron micrographs of protoplasts treated as in *a* and *b* respectively. Os = osmiophilic (lipid) extrusions, Va = vacuoles, V = endocytotic vesicles. Va regions correspond to the extrusions in the SEM micrograph *b*. They remain attached to the cytoplasm (shown by arrows) and are reincorporated into the membrane when the osmotic stress is reduced. The scale markers represent 5 μm. Reproduced from Steponkus *et al.* (1983).

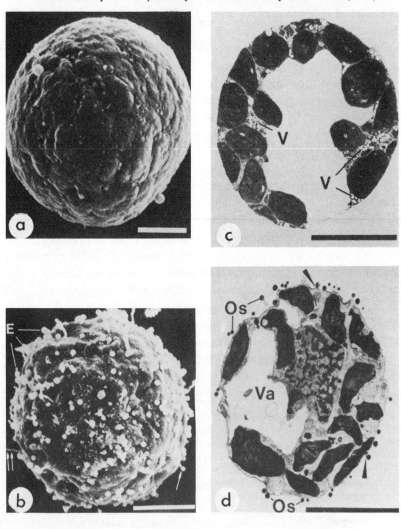

able to reincorporate the lipid bodies into the plasma membrane, so restoring its original appearance. In no case has a folding or deformation of the membrane been observed; the protoplasts do not lose their spherical shape. In contradiction to the conclusions of Williams *et al.* (1981), these observations suggest that the resting tension of the plasma membrane is unaffected by hyperosmotic stress (Wolfe & Steponkus, 1983). The above results are summmarized in Fig. 7.10 where the responses of tender and hardened rye protoplasts to hyperosmotic stress are compared.

7.6   Ice nucleating agents
        Just as antifreeze peptides promote undercooling, presumably by perturbing the interface between the embryonic nucleus and the undercooled aqueous phase, so other types of compounds have been identified whose natural function appears to be the facilitation of ice nucleation *in vivo*. Mention has already been made of nucleation catalysts of biogenic origin (see Chapter 2), but a distinction must now be drawn between biological structures which happen to be able to act in such a manner and those which are synthesized specifically to minimize undercooling of tissue fluids during periods of freezing stress.

It is to be expected that most, if not all, biological cells contain supramolecular structures of various types which could become active as nucleating catalysts at some temperature above the homogeneous nucleation temperature of the cytoplasmic fluid (Mathias, Franks & Trafford, 1984). Such catalytic structures might be of lipid, protein or carbohydrate origin. The properties which render such a structure active as a nucleating catalyst are thought to include its dimensions (radius of curvature), its wettability by ice and water respectively and its degree of molecular symmetry (Fletcher, 1970). In the case of human red blood cells, for instance, internal nucleation of ice appears to be catalysed by the plasma membrane, but in a very ineffective manner: the cells can be undercooled to almost -39 °C before they freeze spontaneously (Franks *et al.*, 1983). Even yeasts and cultured plant cells readily undercool to below -30 °C before some intrinsic structure is able to initiate ice nucleation within the cell. The position is different with some bacteria. The most celebrated biogenic ice nucleating catalyst has been isolated from *Pseudomonas syringae*, a microorganism commonly found on plant leaves (Schnell & Vali, 1972). In its nucleating potency it far surpasses the more conventionally studied silver iodide. It appears that the catalytic site consists of a proteinaceous multisubunit structure. By the removal of the so-called ice nucleating gene it has been possible to produce a mutant (the ice-minus mutant) which is inactive as a freezing catalyst (Lindow, 1983).[†] The capability of *Ps. syringae* and other bacteria to catalyse the nucleation of ice is probably quite fortuitous and can hardly be related to the physiological functioning of such microorganisms. It is reminiscent of the ability of plant seed lectins to agglutinate red blood cells, an ability which must be quite irrelevant to the as yet unidentified biological function of lectins. On the other hand, there are several cases on record of ice nucleation by biogenic catalysts, where such a process is directly linked to the ability of the host organism to survive at subfreezing temperatures. An example is provided by the alpine plant *Lobelia teleki* which grows on the slopes of Mount Kenya at temperatures which fluctuate daily over the range -10 to $+10$ °C (Krog, Zachariassen, Larsen & Smidsrød, , 1979). The inflorescence of this plant contains a potent ice nucleation catalyst which completely suppresses undercooling. When the temperature of the tissue fluids falls to the equilibrium freezing point, -0.5 °C, freezing occurs and the latent heat released prevents further falls in the temperature of the plant during the night. The catalyst, believed to be of carbohydrate origin, has been shown to be active *in vitro*: it can completely inhibit the undercooling of microdroplets of saline solution.

Ice nucleating catalysts have also been identified in overwintering insects (Duman & Patterson, 1978; Zachariassen, 1980). Their function is presumably to prevent undercooling of the extracellular fluid. If freezing can be induced close to the equilibrium freezing point, then the resulting transmembrane osmotic stress is applied gradually as the temperature falls. Provided that the cell is kept in osmotic equilibrium with its surroundings, the likelihood of intracellular ice crystallization is minimized. Freezing studies on the isolated haemolymph of the hornet *Vespula maculata* have shown that, despite a high glycerol concentration, the haemolymph undercooling point lies only 2 deg below the equilibrium freezing point. The catalytic effect can be destroyed by heating the haemolymph to 100 °C or

---

† This piece of genetic engineering has led to curious developments, a typical example of what happens when scientific results become mixed up with politics and the popular news media. Applications to the US National Institute of Health for permission to use the mutant in a field trial in California produced threats of law suits by various pressure groups (David, 1983). In the meantime the issue was picked up by the media and various garbled versions of the story have appeared on the 'Science' pages of newspapers. A director of a Biotechnology company has been quoted in the *Guardian* as claiming that his company would now produce the wild bacterium in large quantities for the purpose of helping winter sports resorts provide sufficient snow.

A confused and confusing report of the ice-minus mutant story has also appeared on the BBC television programme *Tomorrow's World* and we have surely not heard the last of it yet.

by the action of a proteolytic enzyme, but not by dialysis. It is concluded that the nucleating substance is a protein. The catalyst appears to accumulate in the haemolymph during the autumn, because the ability of the haemolymph to undercool decreases during the period August to December.

Here again, it is not strictly correct to describe such behaviour as freeze tolerance, because by facilitating and optimizing extracellular freezing the organisms are able to prevent intracellular freezing which would be lethal. A better way to distinguish between the different responses to subfreezing temperature would be in terms of total freeze avoidance (deep undercooling) and intracellular freeze avoidance through extracellular freezing and (controlled) osmotic dehydration.

8

Cryobiology: the laboratory preservation of cells, tissues and organs

8.1   Objectives

In this chapter we return to the problems first posed in Chapter 5, albeit with a different emphasis. The long term preservation of labile material of biological origin has long been felt to be a desirable objective, and the use of low temperatures to achieve such an end seems the obvious route to follow. The general objectives of such studies fall naturally into several classes. Pride of place has always been given to clinical cryopreservation, the ability to store in a recoverable state a variety of spare parts required by the human body. They cover a wide range, from blood components to whole organs, such as are used for transplantation. Potentially of much greater importance to humanity, if not to those individuals whose tasks include the apportioning of research and development funds, is the preservation of variety in plant genetic stocks which, over past decades, have been rapidly depleted by natural causes, by industrial development and by the plant breeder, working in conjunction with commercial agricultural interests. The third objective follows directly from the rapid developments in biotechnology: the application of cell and tissue culture methods is proliferating at a fast pace. The techniques are labour and capital intensive and subject to problems, such as the probability of genetic drift. An ability to maintain cultures over extended periods in a state of suspended animation is therefore a highly desirable commercial objective. It is not the purpose of this book to present an economic case for the pursuit of cryobiology, but even a layman will appreciate the enormous potential of reliable low temperature storage techniques, should they ever become available.

The history of cryobiology as a scientific discipline is curious. After an early fortuitous breakthrough which gave promise of quick successes, it was subsequently felt to be unnecessary to probe the basic problems posed by freezing live material, since success was just around the corner. In the event this proved not to be so, and for two decades little progress was made, most investigators limiting themselves to insignificant changes in the freezing protocols used. The few studies published during that period which actually produced a new understanding did not receive wide publicity and their significance was hardly appreciated by the practitioners of freeze preservation. The discipline was the preserve of the medical profession with emphasis on clinical techniques. Claims of successes were published but could not be confirmed and, with hindsight, there seems little sense of direction in the published research of the period.

Matters have changed over the past decade. It is too early to decide whether a better understanding of the basic problems involved is already leading to social or economic benefits, but that such benefits will accrue is now only a matter of time. An analysis of the progress made in cryobiology indicates that it dates from a better appreciation of the basic factors involved in the low temperature properties of water and aqueous systems, such as undercooling, ice nucleation and crystal growth, unfreezable water, glass transitions, freeze concentration, cell membrane permeability, and long term changes during storage at subfreezing temperatures, to name but some. Such knowledge, when combined with the physiological and biochemical parameters which govern recovery and functioning, is likely to produce the success which has for so long eluded cryobiologists.

8.2 Basic problems of cell freezing: the function of chemical cryoprotectants

In Chapter 5 the response of a cell to subfreezing temperatures was briefly discussed and we now return to a more detailed analysis of the processes involved. We have seen that a major injurious factor is freeze concentration of the cell contents. Freeze concentration within the cell can come about by one of two possible mechanisms: during slow freezing the flow of water out of the cell in response to the osmotic gradient will lead to a gradual dehydration of the cell. Fast freezing, on the other hand, severely perturbs the transmembrane equilibrium; the water flow out of the cell is then limited by the membrane permeability. Substantial undercooling will occur, leading to rapid intracellular freezing and the concomitant freeze concentration (Mazur, 1970). Fast and slow freezing are relative terms, depending mainly on the permeability of a given plasma membrane to the outflow of water, in response to the particular temperature gradient imposed by the cooling protocol employed.

Before analysing in more detail the consequences of freezing, let us compare the major difference between freezing in the natural environment

(Chapters 6 and 7) and freezing as performed in the laboratory for purposes of cryopreservation. In the natural environment fast freezing does not occur; the cell (or the organism) can always adapt to the osmotic stress exerted by the extracellular ice, provided, of course, that the cell fluids are not isolated from the extracellular fluid by an impermeable barrier. Such barriers may be part of the phenomenon of deep undercooling, as discussed in Chapter 6. The laboratory cryobiologist, by contrast, has at his disposal the control of cooling and warming rates.

Cold hardening *in vivo* frequently involves the biosynthesis of water soluble compounds which mitigate the damaging effects of freeze concentration. Most of the compounds in question are not able to cross the plasma membrame so that they cannot be used as additives in laboratory cryopreservation. On the other hand, the cryobiologist does have control over the concentration of additives, their rate of addition and the temperature at which they are added, and he can use additives that do not occur *in vivo*. He can also develop artificial 'hardening' protocols, where cells or tissues are subjected to mild osmotic stresses prior to freezing. The temperature of storage is yet another parameter that can be controlled in laboratory cryobiology. Mainly for reasons of history and convenience, $-196\,°C$, the boiling point of nitrogen, has been chosen as the standard storage temperature for the cryopreservation of cells and tissues. Little thought seems to have been given to whether such a low storage temperature is necessary or desirable.

In summary, the cryobiologist has at his disposal several controllable variables to aid him in minimizing the two damaging factors: osmotic cell dehydration and intracellular freezing. In order to establish the limits of these factors, Mazur derived quantitative relationships between degree of freeze concentration, temperature and cooling rate by considering the cell as a perfect osmometer, permeable only to water, and by assuming that the protoplasm obeys Raoult's law and that the plasma membrane surface area is not altered during freezing (Mazur, 1970). Over the years following Mazur's original analysis (Mazur, 1966) various refinements have been proposed, mainly to facilitate the solutions of the various differential equations by numerical methods and to correct for deviations from the simple ideal solution model. The general principles, first examined by Mazur, still apply and the brief analysis given in the following paragraphs is based largely on his treatment.

The thermodynamic aspect of freezing relates to the activity difference between ice and undercooled water at a common temperature. For an ideal solution, where vapour pressure can be set equal to activity, and for low degrees of undercooling,

$$\mathrm{d}\ln(p_w/p_i)/\mathrm{d}T = -\Delta H_f/RT^2 \qquad (8.1)$$

where p_w and p_i are the vapour pressures of liquid water and ice, and ΔH_f is the molar heat of fusion. Consider now an aqueous solution whose mol fraction of water is x_w, placed inside the cell (protoplasm) and outside the cell. When the extracellular medium freezes, eqn (8.1) has to be modified to include the change in concentration brought about by freezing:

$$\mathrm{d} \ln (p_w/p_i)/\mathrm{d}T = -\Delta H_f/RT^2 - \mathrm{d} \ln x_w/\mathrm{d}T \tag{8.2}$$

where p_w is now the *partial* vapour pressure of water in the solutions. The second term on the right hand side of eqn (8.2) can be rewritten in terms of V (the volume of water in the cell), v (the molar volume of water) and n (the number of osmoles of solute in the cell), so that

$$\frac{\mathrm{d} \ln (p_w/p_i)}{\mathrm{d}T} = -\frac{\Delta H_f}{RT^2} - \frac{nv}{(V+nv)\,V} \cdot \frac{\mathrm{d}V}{\mathrm{d}T} \tag{8.3}$$

The driving force (free energy) responsible for freeze concentration is provided by the partial vapour pressure ratio on the left hand side of eqn (8.3). The *rate* at which water can leave the cell is a function of the total surface area (A) of the plasma membrane and its permeability (L_p) to water. At any given reference temperature T_r, the rate of loss of water is given by

$$-\frac{\mathrm{d}V}{\mathrm{d}t} = \frac{L_p\,ART_r}{v} \ln (p_w/p_i) \tag{8.4}$$

Finally, if the temperature dependence of L_p is known, then the combination of eqns (8.3) and (8.4) yields an expression for the volume of cell water as a function of temperature. Reliable information on the nature of $L_p(T)$ is somewhat scarce, but it is generally assumed that an exponential function of the type

$$L_p = L_{p_r} \exp (b[T - T_r]) \tag{8.5}$$

is satisfactory, where b is the temperature coefficient of L_p in the neighbourhood of the reference temperature T_r, which is usually close to room temperature. Assuming a constant cooling rate, $\mathrm{d}T/\mathrm{d}t = \theta$, we obtain

$$T \exp [b(T_r - T)] \cdot \frac{\mathrm{d}^2V}{\mathrm{d}T^2} - \left[(bT+1) \exp \{b(T_r - T)\} \right.$$
$$\left. -\frac{ARL_{p_r} \cdot n}{\theta(V+nv)} \cdot \frac{T^2}{V} \right] \frac{\mathrm{d}V}{\mathrm{d}T} = \frac{\Delta H_f\,AL_{p_r}}{\theta v} \tag{8.6}$$

According to Mazur, the most serious source of error arises from a lack of knowledge of b. Equation (8.6) has been solved numerically for several cell types with different permeabilities. Figure 8.1 illustrates two extreme cases: human erythrocytes and yeast cells. The broken line represents cooling under conditions of osmotic equilibrium between the cell and the freezing extracellular fluid, i.e. infinitely slow cooling. At finite cooling rates the protoplasm will be subject to undercooling, the degree of undercooling being given by the temperature difference between the actual $V(T)$ curve and the equilibrium curve. Experience shows that the maximum degree of undercooling that can be tolerated is approximately 15°. The membrane then appears to lose its ability to block ice seeding from the extracellular environment (Mazur, 1953).

Figure 8.1a also includes $V(T)$ estimates arrived at with the aid of a rather more complex model of the water permeability L_p and different values for the cell surface area (Silvares *et al.*, 1975). $V(T)$ is seen to be extremely sensitive to these parameters. On the other hand, making allowances for non-ideal solution behaviour and lifting the restriction of a constant, temperature invariant ΔH_f hardly affects $V(T)$.

A comparison of the responses shown by yeast and erythrocytes to cooling shows that a cooling rate of 1000 deg min^{-1} can be considered slow for erythrocytes, but that a corresponding $V(T)$ behaviour is observed for yeast cells cooled at 10 deg min^{-1}. The difference reflects the high permeability of the red cell membrane and its large surface area : volume ratio. In practice, therefore, cooling rates must be chosen such that intracellular freezing is avoided, while at the same time osmotic dehydration of the protoplasm is minimized. For each cell type there exists a cooling rate where these conditions are met optimally. However, the degree of cell survival, even under such optimum cooling conditions, is so low that freezing in the absence of a cryoprotectant additive is not a practical proposition.

Prior to 1940 only isolated low temperature preservation studies on cells had been reported; the first work on red cell freezing and the effects of

Fig. 8.1. Calculated fraction of water remaining in (*a*) human erythrocytes and (*b*) yeast cells, as they are cooled to various temperatures at the indicated rates (deg min^{-1}); according to Mazur (1970). The dotted lines represent equilibrium volumes (slow cooling) and the broken lines are the $V(T)$ curves based on the model by Silvares *et al.* (1975).

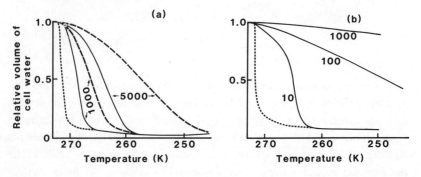

additives on survival appears to date from 1941 (Woodcock, Thistle, Cook & Gibbon, 1941). Ironically the authors discarded the idea of liquid air as a suitable storage medium because 'considerable difficulty would be experienced by providing the requisite liquid temperatures for storage.' Nowadays liquid nitrogen is the conventionally used material. It is difficult to understand how this work came to be overlooked and forgotten in the middle of a war, when news of effective methods for preserving blood should have alerted at least the army medical authorities. In the event, cryopreservation was 'rediscovered' in 1949, when the protective action of glycerol was reported (Polge, Smith & Parkes, 1949). The following years witnessed increasing activity, mainly devoted to the screening of potential cryoadditives and the effects of glycerol on different cell types.

The next major advance occurred when Lovelock & Bishop (1959) reported the efficacy of dimethyl sulphoxide (Me_2SO) as a cryoprotectant for red cells. Their short publication provides no clue why Me_2SO was chosen, except that they state that an effective protectant must be non-toxic, possess a low molecular mass and therefore a high water solubility (the one does not seem to follow from the other) and must be able to penetrate the plasma membrane. In the case of bovine red blood cells Me_2SO fulfils all three conditions. This is not the case for glycerol which is unable to penetrate the plasma membrane of this type of cell. Since the early pioneering days multifarious studies have been conducted, most of them relying on the use of glycerol or Me_2SO. The freezing of many different cell and tissue types has been described with diverse cooling and warming protocols, storage temperatures and cryoprotectant

Fig. 8.2. Generalized interrelationships of cooling rate, cryoprotectant concentration and survival rate. The optimum cooling rate shifts to lower values with increasing cryoprotectant concentration.

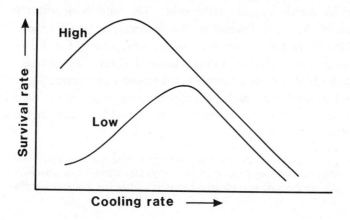

concentrations (Meryman, 1966; Smith, 1970; Ashwood-Smith & Farrant, 1980). Most of the results have been reported in terms of the combined effects of cryprotectant concentration and cooling rate on cell survival, as shown schematically in Fig. 8.2. A general finding has been that an increase in the cryoprotectant concentration produces an improved survival rate at a lower cooling rate.

Although a general qualitative understanding of the observed phenomena has gradually emerged, cryobiology remained a recipe science until quite recently. This is graphically illustrated by recent reviews of the state of the art. Thus, in a chapter on 'general observations on cell preservation', successive sections are headed 'Rapid thawing generally improves survival' and 'Slow thawing is sometimes better than rapid thawing' (Farrant, 1980). Beyond the basic truths embodied in Fig. 8.2, there is as yet little agreement on optimum preservation techniques or the reasons why some techniques work and others do not.[†]

Despite these drawbacks cryopreservation technology has developed, and a limited variety of cell types and organelles can now be stored for extended periods. Tissue and blood banks have been established in many parts of the world and the accent is now on the development of techniques for the long term storage of whole organs, a daunting task indeed (Jacobsen & Pegg, 1984).

8.3   Mechanisms of cryoprotectant action

Given that it is possible to reduce cryoinjury with the aid of certain additives, we then need to inquire into their mode of action and to derive some quantitative relationships between the additive concentration, the various experimental variables and the degree of freeze/thaw survival. Here again Mazur has laid the foundations (Mazur, Miller & Leibo, 1974). Three features are mainly responsible for the reduction of freeze damage afforded by certain organic compounds. The most basic effect is the reduction of the water content of the specimen to be preserved. This reduces the freeze concentration factor, as illustrated by Fig. 8.3 for the system water–NaCl–glycerol. In the absence of glycerol, an isotonic saline solution (0.154 M) suffers a tenfold concentration on freezing to $-6\,°C$. Figure 8.3 shows that the higher the concentration of glycerol in the initial mixture, the lower is the concentration of NaCl at any subfreezing temperature.

[†] A major weakness of cryobiology, as currently practised, is that published techniques sometimes appear to work only in the author's own laboratory with little collaborative or confirmative work to substantiate protocols and claims.

The second factor is the permeability of the cell membrane to the organic cryoadditive. There is little fundamental understanding of the physico-chemical factors which render certain membranes permeable to some organic compounds. For instance, while the human erythrocyte membrane is readily permeable to glycerol, bovine erythrocyte membranes are hardly permeable to this compound and most plant cell membranes are completely impermeable. On the other hand, Me_2SO appears to penetrate all (or nearly all) membranes, and, largely for this reason, has become established as the universal cryoprotectant. Apart from the conventional, penetrating cryoprotectants, other organic compounds also appear to be able to afford some protection against freeze/thaw injury, although they act only in the extracellular fluid and, therefore, play a part in the osmotic dehydration of the cell. Such protection is identical to that discussed in the previous chapter under the heading of freeze tolerance. Compounds which have been studied for their possible cryoprotective action are mainly those which are found in cold hardened organisms: sugars, sugar alcohols, proline and betaines. Cryoprotective action has also been claimed for some water

Fig. 8.3. NaCl concentration factors (1 = 0.85% by weight) *versus* temperature for several weight % ratios of glycerol:NaCl. Data from Farrant (1966).

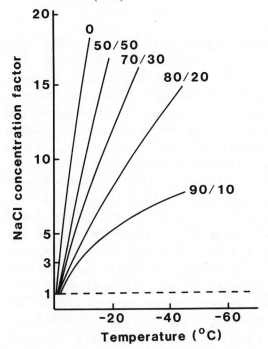

soluble polymers, mainly polyvinyl pyrrolidone, hydroxyethyl starch, polyethylene glycol and dextran. The mode of action of these polymers is even more obscure than that of the natural freeze protecting substances (Meryman, 1974; Echlin *et al.*, 1977); it may be related to their inability to crystallize from aqueous solution. Instead they form supersaturated solutions of very high viscosity, rendering an appreciable fraction of water unfreezable. This aspect of aqueous solutions is treated in the next chapter.

A quantitative description of the changing composition of the cell interior with temperature and time is thus seen to require knowledge of the heat and mass transfer that occur during freezing, with water leaving the cell and the cryoadditive entering, but at a different rate. During this process the concentrations of all membrane impermeable substances undergo changes, as may also the volume and shape of the cell. The development of a credible model embracing all these simultaneous processes is a daunting task, even if the actual solving of the differential equations involved (by numerical methods) has become easier than it used to be. The following account is based largely on the work of Diller and his colleagues, who have combined the physical aspects of thermodynamics and kinetics with computer simulation methods to obtain several quantitative relationships between the various parameters which determine the response of a cell to freezing in the presence of a cryoptotectant – glycerol in this case (Diller & Lynch, 1983, 1984*a*, *b*).

Basic to any such analysis is a knowledge of the phase equilibria that exist in the ternary system water–glycerol–NaCl, it being assumed here that NaCl is the only other constituent present in the cell. At this stage it is also assumed that ice is the only crystallizing component and that phase changes take place under equilibrium conditions. The liquidus surface of such a ternary system defines the composition of the liquid phase in equilibrium with ice at any given temperature. In order to simplify matters it is helpful to study so-called isoplethal sections of the surface; these are sections for which the weight % ratio glycerol:NaCl is constant while only ice crystallizes (Shepard, Goldston & Cocks, 1976). The usefulness of this approach is that it converts the ternary system into a quasi-binary one of water + solute. A set of such liquidus curves is shown in Fig. 8.4 with solute ratios ranging from zero to infinity. The curves end at the respective quasi-binary eutectic temperatures where NaCl crystallizes. They are depressed by the addition of glycerol. The effect of glycerol on the freezing point depression is demonstrated by the curvatures of the coexistence lines.

The data in Fig. 8.4 can now be combined and represented as a projection of the ternary liquidus surface onto the ternary composition triangle. This is shown in Fig. 8.5 for the water rich corner of the phase diagram. The area of interest is bounded on the right by the eutectic trough where NaCl . 2H$_2$O crystallizes and at the top by the eutectic trough where glycerol would crystallize if equilibrium conditions could be assured. Ternary equilibrium data make possible the estimation of the NaCl concentration factors shown previously in Fig. 8.3. A comparison of the experimental results with those estimated from ideal solution calculations shows that glycerol is actually more effective in 'diluting' NaCl than such calculations had indicated. On the other hand, Me$_2$SO is even more effective than glycerol in protecting a cell against salt injury. The information contained in Figs. 8.3 and 8.5 makes possible a prediction of the temperature at which the NaCl concentration reaches the threshold value for cryoinjury through haemolysis. Thus, the isotonic NaCl concentration for human erythrocytes is 0.85 weight % and haemolysis is first observed at 4.41 weight %. Calculations based on ideal solution behaviour predict that at 5% glycerol levels this occurs at −7 °C, whereas phase diagram data show that the actual temperature is −16.5 °C.

Turning now to the dynamic aspects of freezing and cryoprotection, equations have to be established for the complex counterflow of water and glycerol at finite cooling rates. They follow from eqn (8.6) which applies

Fig. 8.4. Intersections of isoplethal sections with the liquidus surface of the water–NaCl–glycerol system, for different glycerol:NaCl weight ratios. Horizontal bars represent eutectic temperatures and the broken line denotes a supersaturated mixture. After Shepard, Goldston & Cocks (1976).

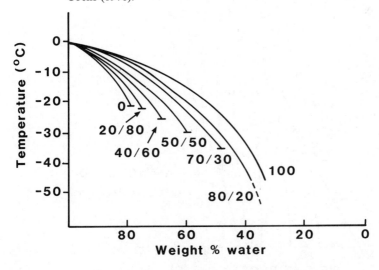

to the osmotic flow of water from the cells in the absence of glycerol. The two equations which describe the intracellular concentration of water and glycerol, respectively, are:

$$\frac{dn_w}{dT} = \frac{AL_{p_r}RT}{\theta v} \exp[-\Delta E/RT]\left[\Delta c_e + \left(\sigma + v_g\frac{\omega_r}{L_{p_r}}\Delta c_g\right)\right] \quad (8.7)$$

$$\frac{dn_g}{dT} = \frac{AL_{p_r}RT}{\theta} \exp[-\Delta E/RT]\left[(1-\sigma)c_g^{av}\Delta c_e\right.$$

$$\left. + \left\{\sigma(1-\sigma)c_g^{av} - \frac{\omega}{L_{p_r}}\right\}\Delta c_g\right] \quad (8.8)$$

where the following notation has been employed: n_w and n_g are the number of mols of water and glycerol respectively, c_e is the concentration (mol l^{-1}) of electrolyte, σ is the reflection coefficient and ω is the membrane

Fig. 8.5. Projection of the water–NaCl–glycerol liquidus surface for water-rich mixtures. The solid line represents the ice–NaCl.2H$_2$O eutectic trough. The isothems are obtained from the data in Fig. 8.3. After Shepard, Goldston & Cocks (1976).

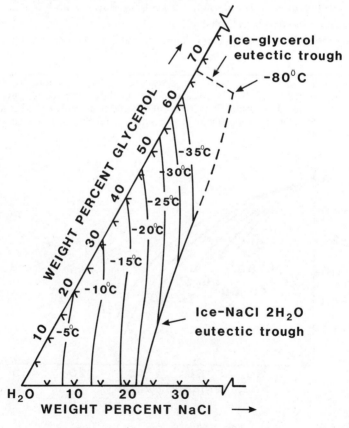

permeability towards glycerol; v_g is the molar volume of glycerol and the superscript av denotes an average value. The other symbols have already been defined.

The numerical solution of the two simultaneous equations has been achieved and the material fluxes have been calculated for various glycerol concentrations (0.4–3.0 M) and a wide range of cooling rates (55–5000 deg min^{-1}). The most significant result is a strong interaction between θ and the glycerol concentration in determining the osmotic response of the cell. The equilibrium phase behaviour, the osmotic stress and the net volume flux are also important.

Assuming that the most injurious factor determining cell survival is the transient increase in the salt concentration during freezing, then it is important to obtain data on how c_e changes with glycerol content and θ. This is shown respectively in Figs. 8.6 and 8.7. The model indicates that the transient salt concentration strongly depends on the glycerol concentration, mainly because of the diluent action of the cryoprotectant, but there are subsidiary effects due to the freezing point depression of glycerol and its exclusion from the (extracellular) ice phase. This latter

Fig. 8.6. Transient intracellular salt concentration during freezing as a function of the initial glycerol concentration. Cooling rate: 100 deg min^{-1}. Reproduced from Diller & Lynch (1984*a*).

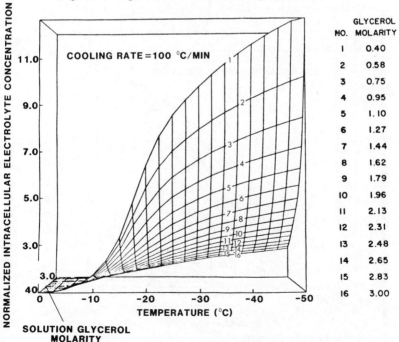

effect reduces the driving force for cell dehydration. Figure 8.7 graphically illustrates how slow cooling leads to a build-up of salt which may well exceed the lethal threshold value. Here again, the actual numbers indicated may not correspond exactly with the experimentally determined survival rates, because the numerical solutions of eqns (8.7) and (8.8) sensitively depend on the actual values used for v, A, L_p, ω, $\varDelta E$ and σ. However, the trends portrayed by the surfaces in Figs. 8.6 and 8.7 are of great interest and importance to practical cryopreservation technology.

To complete the picture of the cell during freezing it is necessary to consider the transient osmotic stress produced by differential fluxes of water and glycerol and the net rate of increase in the intracellular glycerol concentration. The rate of flow of water out of the cell for a 2 M glycerol solution is shown in Fig. 8.8. The general trends are as expected, but the existence of a trough in the surface defining maximum water flux at each isotherm is a novel feature. The practical consequence is that at low temperatures the cell may exhibit an unanticipated response to cooling: as the cooling rate is increased, so is the rate of loss of water.

Figure 8.9 illustrates the glycerol flux for the same system under the same

Fig. 8.7. Transient intracellular salt concentration during freezing as a function of the cooling rate. Glycerol concentration: 2.0 M. Reproduced from Diller & Lynch (1984a).

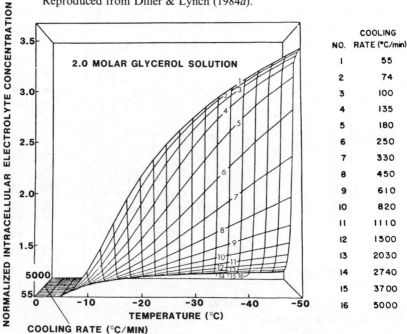

Fig. 8.8. Cell water flux during freezing as a function of cooling rate and an initial 2.0 M glycerol concentration. Cooling rates as in Fig. 8.7. Reproduced from Diller and Lynch (1984*b*).

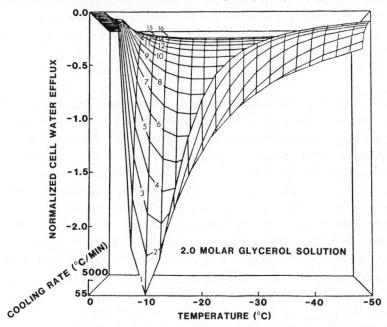

Fig. 8.9. Glycerol flux during freezing as a function of cooling rate and an initial 2.0 M glycerol concentration. Cooling rates as in Fig. 8.7. Reproduced from Diller & Lynch (1984*b*).

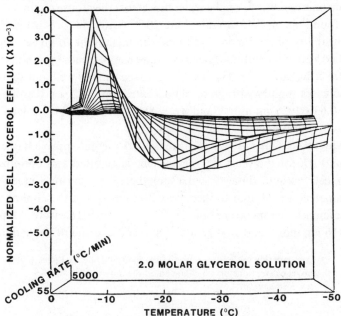

conditions. The most striking feature is the flux reversal which reaches a maximum at cooling rates of 135 deg min^{-1}. This observation has important implications for the way in which cells are incubated in the cryoprotectant mixture. It is standard practice to pre-equilibrate cells prior to freezing, but the glycerol flux data suggest that the higher is the initial glycerol concentration *inside* the cell, the more pronounced will be its depletion during the course of freezing.

8.4   Cell membrane permeability and its temperature dependence

The success and usefulness of all models for the prediction of freeze injury and its prevention depend critically on a knowledge of the properties of cell membranes, in particular the membrane permeability to various substances over a range of temperatures and cooling rates. It can also not be taken for granted that the membrane preserves its barrier properties under conditions of stress, whether of a mechanical or osmotic nature. Indeed, where such barrier properties rely on the effective functioning of ATP-dependent pumps to prevent leakage of intracellular components, it is to be expected that low temperature alone, even without the additional stress due to freezing, might lead to irreversible injury by leakage (cold shock symptoms).

The direct measurement of permeability under conditions of freezing is extremely difficult, so that preference is given to isothermal measurements in which the freezing stress is simulated by a purely osmotic stress at an ambient temperature. Thus, Leibo (1980) has reported on the effects of hypertonic conditions on the water flux across the mouse ovum membrane. He found $L_{p_r} = 0.43 \, \mu$m min^{-1} at 20 °C with $\Delta E = 54$ kJ mol^{-1}. The permeability of the bovine red blood cell membrane to glycerol has been studied both by hypotonic haemolysis and osmotic shock methods (Mazur, Miller & Leibo, 1974). The results agree reasonably well and suggest that ω increases slightly with glycerol concentration. In 2 M glycerol at 10 °C, $\omega = 0.63 \times 10^{-2} \, \mu$m min^{-1}, and $\Delta E = 88$ kJ mol^{-1}, i.e. ω is halved for a 5° reduction in the temperature.[†]

Schwartz & Diller (1983*a, b*) have used a different approach to calculate L_p and ΔE for the yeast cell membrane. They fitted and optimized their own cell volume data to values predicted by various models. Both parameters were found to increase with increases in the cooling rate. At a reference temperature of 20 °C, L_p and ΔE were respectively 0.0116 μm min^{-1} and 19.4 kJ mol^{-1} at a cooling rate of 9 deg min^{-1}; at a

[†] Compare this value for ω with the one used by Diller & Lynch (1984*a, b*) in the model calculations on which are based the data in Figs. 8.6–8.9:
$\omega = 3.36 \, \mu$m min^{-1}.

cooling rate of 35 deg min^{-1} the corresponding values were 2.11 μm min^{-1} and 101 kJ mol^{-1}. The method has not yet been applied to the estimation of the permeability to glycerol.

There are no reliable quantitative estimates of the corrections to be applied in model freezing calculations for transmembrane fluxes of components other than water and cryoprotectant. That such fluxes occur is well established (Mazur, Leibo & Miller, 1974). For instance, as the erythrocyte shrinks during freezing, so the charge on the haemoglobin is reduced, leading to an inflow of Cl$^-$ ions to neutralize the cations released. It is also known from simple electrochemistry that osmotic coefficients of electrolytes at high concentrations exceed unity, implying that the calculated $V(t)$ values underestimate the real values.

8.5   Membrane leakage and cell injury

The above discussion leads to the conclusion that the unique ability of biological membranes to act as selective barriers is of prime importance in freeze/thaw survival. This is not to say that a temporary or partial loss of this ability is in every case lethal. The differentiation between lethal and nonlethal leakage has been studied by Griffiths & Beldon (1978) with the aid of radio labelled markers to monitor the loss of various cell constituents during freezing and thawing. Table 8.1 summarizes the markers used in their studies. The cells (human diploid cells and hamster ovary cells) had been treated with Me$_2$SO and were cooled at different rates, the loss of radio labelled compounds being monitored as a function of temperature and related to cell recovery after thawing. Although the hamster ovary cells were generally more resistant to freezing than the human diploid cells, there are common trends in the results of

Table 8.1. *Radiomarkers used by Griffiths & Beldon (1978) to simulate the leakage of cellular constituents during cooling and freezing*

Marker	Cell constituent
Chromate	Protein (non-specific)
Fucose Cholesterol Palmitic acid	Membrane material
Uridine Proline	Structural proteins
Rubidium	Intracellular ion pool

the leakage experiments; they are summarized in Fig. 8.10. A general leakage of soluble cell components takes place even during cooling from 20 to 0 °C. This is not injurious and may just reflect a Me_2SO effect on membrane permeability and/or the incidence of cold shock. Cell death can be correlated (irrespective of cooling rate) with the loss of proteins, nucleotides and the massive leakage of ions. Membrane markers (fucose, palmitic acid and cholesterol) are only released at quite low temperatures, where recovery is in any case minimal.

The observation that membrane precursors are not removed at temperatures > -40 °C suggests that the membrane itself is not irreversibly damaged under such conditions, although it could be argued that a membrane could be subject to structural changes which would allow gross leakage of ions without protein denaturation or loss of specific membrane components. Significantly, cell death was correlated by the authors with loss of cytoplasmic material rather than membrane markers. As expected, rubidium (as marker for all small ions) proved to be most sensitive to cold shock and freezing stress. On the other hand, loss of ions is not necessarily lethal, but is often temperature reversible.

The complex pattern of selective leakage of cell components under conditions of chill, osmotic imbalance and freezing stress emphasizes that the simple model of the cell as an ideal osmometer provides at best an

Fig. 8.10. Leakage of radio-labelled markers after cooling cells at different rates and thawing from different temperatures between 0 and -196 °C. Also shown (++++++) is the fractional recovery of the cell samples. Leakage rates are represented as follows: severe (——), moderate (——) and slight (– – – –). After Griffiths & Beldon (1978).

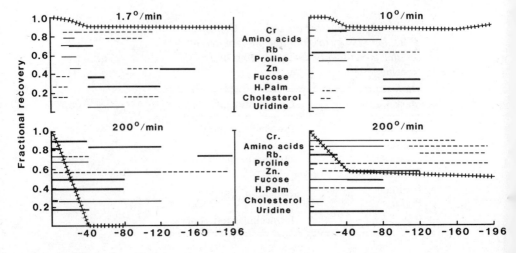

approximate and semi-quantitative description of the cell response to freezing. Considerable refinements to the model are required before calculations, such as those described in the previous section, can provide freeze/thaw protocols that will maximize cell recovery rates.

8.6   Tissue and organ preservation: future prospects

A full discussion of tissue and organ preservation technology is out of place in this book where the primary concern is with the molecular aspects of cold and freezing. However, the driving force for much of the work described in this book, and nearly all the work described in this chapter, is the hope that it will provide us with the capability of preserving tissues, organs and even whole organisms. It is therefore appropriate to assess briefly the current state-of-the-art and to speculate on future progress.

Methods which have been developed for the preservation of single cells and small cell clusters have invariably proved unsuccessful when applied to mature organs. Patient methodological research has isolated some of the problems, but has not yet produced solutions. The geometry of organs and the different permeabilities associated with the various membrane systems set limits to the fluxes of cryoprotectant substances, thus leading to additional concentration gradients. The same heterogeneity also gives rise to additional temperature gradients during cooling and rewarming, making it almost impossible to achieve uniform cooling rates throughout the organ. Another problem concerns the ratio of actual volume occupied by cells to extracellular space. This ratio is very high in organs, and there is evidence that a factor in slow freezing injury may be related to increasingly higher fractions of the volume being occupied by cells (Ashwood-Smith, 1980). The vascular system, the proper functioning of which is a prerequisite for successful preservation, is particularly susceptible to freezing damage. In cell suspensions the growth of extracellular ice is not usually a damaging factor, but it may be so in vascular tissue. The amount of such damage might be limited if it were possible to control the location, amount and morphology of extracellular ice in the tissue. Novel approaches include attempts to prevent, or at least reduce, freezing with the aid of rapid cooling protocols in combination with high levels of cryoprotectant or high hydrostatic pressures (Fahy *et al.*, 1984). These are the conditions which promote vitrification, but aqueous systems require high concentrations of additives, high pressures and high cooling rates – perhaps higher than can be achieved in practice – for even partial vitrification. In order to minimize the toxic effects produced by some cryoadditives, experiments are in progress with cryoprotectant mixtures, but it is too early to assess whether the extreme measures required for vitrification will lead to successful methods for cryopreservation of organs.

Perhaps the outstanding success in preservation technology to date is the development of freeze/thaw protocols for mammalian embryos. Survival rates of up to 80% have been reported and have led to the application of embryo storage as a viable technique in mammalian genetics and in animal breeding programmes (Whittingham, 1980). The basic principles do not differ from those described for single cells, with Me_2SO concentration and cooling/warming rates as the critical variables.

Other aspects of cryobiology include the preservation of plant tissue, spermatozoa, microorganisms and insects. Although isolated successes have been reported, they have been fortuitous. We are still largely ignorant of the factors that make for success with one type of cell or tissue and failure with a very similar system. Progress is also hampered by the publication of dubious claims of successful preservation techniques which are hard to substantiate because of the lack of the necessary details in the published accounts (Bajaj, 1976).

Cryobiology has nevertheless established itself as a discipline in its own right and it is to be hoped that future investigators will be able to build logically on the quantification of the basic principles that have already been established.

9

The technology of metastable water

9.1   Water at subzero temperatures: thermodynamics *versus* kinetics

A sound knowledge of solid–liquid phase relationships in multi-component mixtures is a *sine qua non* for an intelligent exploitation of the low temperature properties of aqueous systems, but that is not to say that this knowledge alone is sufficient for the design of commercially or biologically useful processes. By and large, equilibrium considerations are of little use in industrial processes or final products, so that it is left to the experience and ingenuity of the processor to achieve a degree of thermo-dynamic instability combined, however, with a high degree of physical and mechanical stability. Ice cream as a manufactured low temperature product graphically illustrates this principle, and we shall return to a discussion of ice cream technology in section 9.5. To achieve the desired end it is necessary to process a mixture of ingredients in such a way that the final state does not correspond to the stable equilibrium state or phase behaviour, but that the compound mix can be maintained in its metastable state for sufficiently long periods. What is considered sufficient depends largely on the end use of the product. For instance, in the storage of seeds as a means of preserving germ plasm, such a period is measured in years or decades, whereas in the storage of ice cream it is measured in weeks.

In terms of physico-chemical principles, the compounded system must be prevented from reaching its state of minimum free energy. In practice this is achieved by the interposition of a kinetic activation barrier designed to retard the rate of achievement of equilibrium, as shown schematically in Fig. 9.1. A simple example of this principle is provided by an emulsion: a homogeniser is used to disperse one liquid in another, immiscible liquid phase. Without a kinetic barrier against coalescence, the two phases will tend to separate and achieve the state of lowest free energy in which their interfacial contact area is minimized. The rate of coalescence and phase

167

separation can be reduced by the addition of an emulsifier which provides for a repulsive interaction between colliding droplets and/or by the addition of a stabilizer to increase the viscosity of the continuous phase, thus reducing the collision rate between the dispersed droplets.

In aqueous systems at subfreezing temperatures the stable water phase is ice. Depending on the particular process and system, it may be desirable or essential to inhibit freezing altogether or to let it proceed to a given extent, but to control the type and concentration of ice crystals formed. Both objectives involve the creation of thermodynamically metastable states which must then be 'frozen-in' to achieve the required storage life. In the absence of such stabilization, ice will tend to recrystallize, more ice may form, other components may crystallize and undesirable chemical reactions may occur as a result of increasing freeze concentration. In this chapter we discuss the various aspects of metastable water at subfreezing temperatures and in complex systems: its production, stabilization and some applications. The physical principles involved have already been outlined; they include undercooling, supersaturation, vitrification, crystallization, maturation and melting. We are now concerned with the practical limits set by the physical properties of water and ice and with technological and economic objectives.

9.2   The vitrification of liquid water

For the exploitation of some of the attributes of low temperature systems and techniques it would be highly desirable to circumvent

Fig. 9.1. Free energy diagram showing the relationship between equilibrium and various metastable states for water. The height of the activation barrier to diffusion determines the mechanical stability of the metastable states.

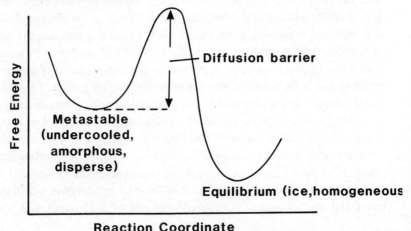

crystallization altogether, mainly because of the deleterious effects of freeze concentration. There have from time to time been claims for the vitrification of water, but usually such reports do not make it clear by what criteria successful vitrification is measured. The following outline describes the parameters which govern the vitrification of an undercooled liquid (Turnbull, 1969; Franks, 1982c). In the first place it must be realized that the definition of a glass is somewhat arbitrary. It is commonly described in terms of a viscosity (say 10^{14} N s m^{-2}). By scaling the viscosity to the molecular diffusion period τ, the glass viscosity corresponds to τ values of the order 10^5 s, corresponding to approximately one day. Since τ corresponds to the time taken by a molecule to diffuse through a molecular distance or perform a molecular rotation, the glass transition viscosity corresponds to a state in which there is practically no diffusive movement at all on a measurable time scale. For a liquid like water, which is very mobile at ordinary temperatures, there must exist a temperature range in which the viscosity rises steeply with decreasing temperature. The Fulcher equation adequately describes the viscosity behaviour of glass-forming liquids:

$$\eta = A \exp[B/(T-T_0)] \tag{9.1}$$

where A, B and T_0 are constants depending on the material. This relationship has already been applied (see eqn (8.5)) to the estimation of membrane permeabilities. When $T_0 = 0$, eqn (9.1) reduces to the well-known Arrhenius equation. As shown in Fig. 3.3, undercooled water exhibits a highly non-Arrhenius type of behaviour. The viscosity can increase rapidly with decreasing temperature either if B is very large (e.g. pure silica) or, if B is not large, then T must approach T_0. For water T_0 is believed to lie in the neighbourhood of 228 K, so that $T_0/T_f = 0.84$ which is large for a molecular liquid. Thus, the viscosity of water would be expected to rise steeply as T approaches 228 K. The transition from a fluid to a glass would then take place over quite a narrow range of temperature. Figure 9.2 compares the viscosity behaviour of silica with those of simple organic glass formers (e.g. isobutyl bromide) and water.

The definition of a glass in terms of the viscosity of an undercooled liquid is perhaps unfortunate because it depends on the method of preparation. It would be more helpful to define the glass in terms of its own properties. This may or may not be possible, but on occasions a glass is incorrectly defined as a crystalline solid with very small grain sizes (Riehle, 1968). Such a duplex mixture of highly ordered and highly disordered material, although having the same X-ray diffraction properties as a glass, should not be so defined.

An ideal glass could be regarded as a solid in internal equilibrium with a definite set of positions about which a molecule can oscillate but without translational symmetry and without a periodic pattern; i.e. a solid with an infinitely large unit cell. In practice a glass approximates to a micro-crystalline solid in which the size of the crystallites is less than five molecular dimensions.

In order to quench a liquid into the glassy state it is essential to control the nucleation and prevent the growth of crystals. This aspect has already been discussed in Chapter 3 (eqn (3.12)). The rate of crystal growth is related to the average elementary diffusive step time τ and the degree of undercooling ΔT. For purposes of comparisons between different types of liquids it is preferable to use the reduced degree of undercooling, $\Delta\theta$, as defined in Chapter 2. Assuming that the viscosity scales with τ, a good working relationship is given by

$$\tau^{-1} = u = k_u \eta^{-1} f(\Delta\theta). \tag{9.2}$$

where k_u is a constant whose numerical value depends on the model used to describe crystallization; for water it can be taken as 10^{23} N m. Equation (9.2) predicts that for $\Delta\theta = 0.1$ and a liquid with the viscosity of water, u is of the order of 10^{10} to 10^{11} molecular spacings per second. On the other hand, for the same $\Delta\theta$, but a viscosity corresponding to that characteristic of T_g, u is only one molecular spacing per 10 to 100 days. Thus, even a highly nucleated glass should resist crystallization at a measurable rate, and this is sometimes observed. In other cases crystal growth is catalysed by the incorporation of impurities which reduce τ in the interfacial region.

Fig. 9.2. The relationship between log (viscosity) and θ^{-1}, the reciprocal reduced temperature ($\theta = T/T_f$) for silica (------), a typical molecular glass former (——) and water (——). An approximate extrapolation suggests that for the vitrification of water $\theta^{-1} \sim 2.0$, i.e. $T_g \sim 136$ K. Adapted from Franks (1982c).

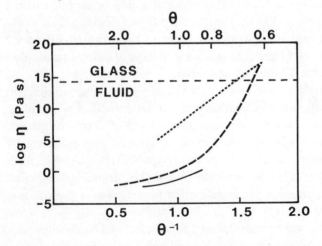

The kinetics of nucleation, as discussed in Chapters 2 and 3, are such that this process can probably not be suppressed in water or aqueous solutions, unless the viscosity is very high or the factors that contribute to G^* (see eqn (2.4)) make this quantity very large. In general the number of critical nuclei dn^* that are formed in a volume v_1 and time dt is

$$dn^* = Jv_1 \, dt \qquad (9.3)$$

For a low viscosity liquid u is so large that it is practically impossible to maintain isothermal conditions. The maximum cooling rate is then limited by the dissipation of the heat of crystallization. To achieve vitrification, nucleation would have to be completely suppressed, so that

$$n^* = v_1 \int_0^t J(T) \, dt < 1 \qquad (9.4)$$

where t is the time in which the liquid is cooled and J is now a function

Fig. 9.3. Estimated $J(\Delta\theta)$ curves for different values of T_f, σ and $\Delta E^{\ddagger}$ for viscous flow. After Muhr (1983). The bold curve is based on the most recent experimental nucleation data and the measured physical properties of undercooled water, as discussed by Franks, Mathias & Trafford (1984).

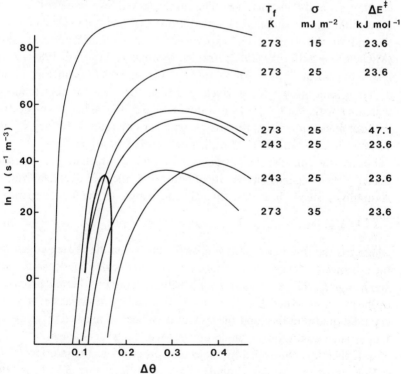

T_f K	σ mJ m^{-2}	$\Delta E^{\ddagger}$ kJ mol^{-1}
273	15	23.6
273	25	23.6
273	25	47.1
243	25	23.6
243	25	23.6
273	35	23.6

of temperature. Equation (9.4) indicates that n^* will be minimized for small volumes, low nucleation rates and high cooling rates.

The minimum time required for the formation of the first nucleus, t_{min}, is given by $t_{min} = i^*\tau$, where i^* is the number of molecules required to form a stable nucleus (i.e. one capable of growth). Although the precise magnitude of i^* for water under different conditions is subject to some uncertainty, the data in Table 2.2 suggest that, as T_h is approached, $10^2 < i^* < 10^3$. Putting $\tau = 10^{-12}$ s, this gives $t_{min} \simeq 10^{-10}$ s. In other words, to avoid nucleation, the liquid has to be cooled to the glass temperature in $< 10^{-10}$ s, equivalent to a cooling rate dT/dt of the order of 10^{12} deg s^{-1}.

The kinetic analysis of nucleation outlined in Chapter 2 (eqn (2.6)) indicates that the most important factor opposing the creation of a stable nucleus is σ, the interfacial free energy between the surface of the growing cluster and the undercooled liquid phase. Unfortunately σ is inaccessible to direct measurement and can only be estimated with the aid of high precision nucleation measurements coupled with the assumption that the stepwise growth model is applicable. Figure 9.3 illustrates the sensitivity of $J(\theta)$ on the assumed values of some of the properties that make up the constants A and B in eqn (2.6a). The curve marked 'experimental' is based on a minimum of simplifying assumptions and has been calculated with the aid of the most recent experimental data on undercooled water (Michelmore & Franks, 1982; Franks, Mathias & Trafford, 1984). $J(\Delta\theta)$ is seen to peak at 2×10^{14} m^{-3} s^{-1} and $\theta = 0.154$. Assuming once again that J and u scale as η^{-1} for a given $\Delta\theta$, then the glass-forming tendency increases with $\theta_g (T_g/T_f)$. For water $\theta_g \simeq 0.5$. To achieve vitrification without nucleation might be possible with cooling rates $> 10^6$ deg s^{-1} and volumes not exceeding 10^{-13} m^3, i.e. droplets of 30 μm radius.

Let us now consider the vitrification of a liquid which already contains nuclei: u then depends on the rate of removal of the heat of crystallization. Assuming a steady heat flow through the sample to the low temperature sink,

$$u = (10^{10} \lambda\beta f)/\eta \cdot \Delta\theta/(1+K) \qquad (9.5)$$

where f is the fraction of lattice growth sites which lie at step edges, λ is the advance of the crystal interface for each molecular condensation step and $\beta = \Delta H_c/RT_f$. K is a parameter which only contains thermal properties of the material. When $K \gg 1$, then u is governed by heat transfer from the crystal–liquid interface and the system is subject to marked recalescence. Under such conditions a single nucleus will cause crystallization. When $K \ll 1$, then u is controlled by molecular processes at the interface.

For most molecular liquids the condition for $K < 1$ is that

$\eta > 0.1$ N s m^{-2}. A liquid with an average nucleus density of 10^9 m^{-3} could be easily vitrified when cooled at 0.1 deg s^{-1}; this is the case for fused silica or glycerol, but not for water. The ease of vitrification is determined by experimental variables such as dT/dt and v_1, but the limits are set by physical properties of the material, among them θ_g, f, the nucleus density, the thermal conductivity and ΔH_c. Most of the above estimates refer to pure molecular liquids. The presence of solutes (impurities) increases the driving force for crystallization, but such an increase in u may well be more than counterbalanced by a reduction in the kinetic coefficients, the net effect being an enhanced ease of vitrification.

The question remains whether it is possible in practice to undercool liquid water into the glassy state, as distinct from microcrystalline ice – a distinction which is not always made (Riehle, 1968). Several reports have appeared claiming successful vitrification of small droplets and thin films (Brüggeller & Mayer, 1980; Mayer & Brüggeller, 1983; Dubochet, Adrian & Vogel, 1983). The criterion of success is the detection of a devitrification transition to cubic ice (Ic) by X-ray or electron diffraction or by scanning calorimetry. Since ice Ic cannot be prepared directly from hexagonal ice (Ih), but only from amorphous ice, its formation during the rewarming of rapidly quenched liquid water substantiates the claim for vitrification, or at least partial vitrification. The devitrification temperature of amorphous ice (whether previously formed by quenching the liquid or the deposition of water vapour onto a cold surface) has been estimated as ~ 151 K and the density of the amorphous form is 0.94 g cm^{-3}, compared to 0.92 g cm^{-3} for ice-Ih at the same temperature.[†]

It might be said that a preoccupation with fine distinctions between vitrification and microcrystallinity is hair splitting, particularly when the dimensions of samples that might be properly vitrified can never exceed a few micrometers. For some technological applications this is probably true, but when the area of observation is itself measured in micrometers (electron microscopy), then such differences become significant, as they also do where biological preservation is involved. Thus, a frozen system, even if highly microcrystalline, is still subject to freeze concentration of all water soluble components, whereas the molecular and ionic distributions in a vitreous system are unchanged from those in the liquid state. Before

[†] The recent studies on rapidly quenched water may eventually help to resolve a long-standing dispute as to whether there exists a continuity of state between amorphous ice (i.e. water vapour condensed onto a solid surface at temperatures in the neighbourhood of 80 K) and undercooled water, or whether these two forms of water are structurally and energetically distinct from each other (Angell, 1982; Sceats & Rice, 1982; Mayer & Brüggeller, 1983).

turning to the applications of metastable water, let us discuss the theory, practice and limits of rapid cooling, as applied to aqueous systems.

9.3   Non-equilibrium cooling and the crystallization of ice

Having made the point that in the technological context equilibrium freezing is usually highly undesirable, we now investigate the practical limits to metastability in the frozen system. The problem is one of relating cooling rate and the temperature of the cryogen to crystal growth and crystal size distribution within a sample. The heat flow balance is obtained by equating the sum of the energies supplied to a system to the sum of the energies stored by, or removed from, the system. For a cylindrical sample of radius R, the cooling rate at any point r (*where* $0 < r < R$) is given by

$$\frac{\partial T}{\partial t} = \frac{\bar{k}}{\rho \bar{C}}\left(\frac{\partial^2 T}{\partial r^2} + \bar{k}^{-1}\frac{\partial \bar{k}}{\partial r}\cdot\frac{\partial T}{\partial r} + r^{-1}\frac{\partial T}{\partial r}\right) + \frac{\Delta H_c}{\rho \bar{C}} \tag{9.6}$$

Here ρ is the density (taken to be constant), $\bar{k}$ is the mean thermal conductivity and $\bar{C}$ the mean specific heat. Equation (9.6) demonstrates that the *actual* cooling rate is determined by the temperature gradient within the (homogeneous) sample and by the warming rate due to the release of heat during crystallization. When the latter exceeds the former, the temperature within the sample will actually rise, as illustrated by the typical cooling curve in Fig. 3.7. Under rapid cooling conditions such a warming up effect (recalescence) may only be transient and incomplete, but it will nevertheless affect the local crystal size distribution. Equation (9.6) can be simplied if the cooling rate is high and the fraction of water crystallized is low:

$$\frac{\partial T}{\partial t} = \frac{\bar{k}}{\rho \bar{C}}\left(\frac{\partial^2 T}{\partial r^2} + r^{-1}\frac{\partial T}{\partial r}\right) \tag{9.7}$$

In this case the cooling rate in *liquid* water is a monotonic function of r^{-1}.

The dimensions of the crystals formed are primarily determined by $J(T)$. According to Riehle, the transition from macroscopic crystals to crystals of nanometer dimensions occurs almost discontinuously at a critical cooling rate and over a limited temperature range; for water under a pressure of 200 MPa this has been estimated as 203–223 K. Equation (9.6) has been solved numerically for water held in a cylindrical vessel of $R = 100\ \mu m$ under a pressure of 200 MPa. Under these conditions the critical cooling rate is of the order of 7×10^3 deg s^{-1}. The results are shown in Fig. 9.4: for $r > 10\ \mu m$ the cooling rate exceeds this value and the fraction of ice formed is < 0.2, the crystal dimensions being of the order of 1 nm; the cooling profile is in agreement with eqn (9.7). However, in the interior of the sample ($r < 10\,\mu$m) cooling is much slower, and substantial recalescence takes place with the temperature rising steeply and a 100% conversion to ice with a mean crystal radius of 2 μm.

With the aid of the relationships in eqn (9.6) it has also been predicted that in a 20% glycerol solution cooling velocities of $> 10^4$ K s^{-1} throughout the sample could be achieved only for samples with $R < 50\,\mu$m.[†]

Just as calculations of the parameters governing the vitrification of water or the achievement of microcrystallinity involve several unknown quantities, so the actual measurement of cooling rates is also subject to uncertainties. The cooling rate, dT/dt, which is central to the various calculations can either be measured directly or estimated by inference. Thus, if it can be shown unambiguously that a sample of water has been quenched into the vitreous state, then we can infer that a cooling rate in excess of 10^6 deg s^{-1} must presumably have been achieved during quenching. This inference is itself based on the validity of the underlying theory. Alternatively, the cooling rate could be measured, but this is beset by problems of a different kind. Thermocouples can only monitor the temperature at their area of

Fig. 9.4. Crystallization rates for different spherical volume elements of water and the corresponding volume fractions (ϕ) of ice formed (broken lines); data obtained by the numerical solution of eqn. (9.6) for water under a pressure of 200 MPa. For details see text. After Franks (1982c).

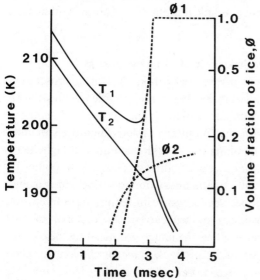

[†] Both the use of glycerol and the application of high pressure serve to retard the rate of crystallization of ice. It is uncertain whether, or how, the calculated values in Fig. 9.4 could be extrapolated for pure water at atmospheric pressure.

contact with the sample being cooled, but we have already seen that the temperature gradients within a sample are not linear, and in the case of recalescence, the temperature profile is extremely complex (see Fig. 9.4). Measured cooling rates can also be subject to several thermal artefacts: the geometry and dimensions of the sample holder, the method of cooling and effects arising from the thermocouples themselves (Costello, 1980).

Over the past several years important developments have been achieved in rapid cooling techniques of aqueous samples (Echlin, 1985). Consideration must be given to the cryogen used and to the method by which the sample is brought into contact with the cold sink. The choice of cryogen is determined by its thermal properties, such as boiling and freezing points and thermal conductivity. Substances which find general application are liquid nitrogen at its freezing point (77 K), liquid propane or ethane and liquid helium. The sample can either be plunged into the liquid cryogen at a controlled rate, or a cryogen jet can be made to impinge onto the sample. Where the sample can be dispersed into small droplets by forcing it through a nozzle, it can be spray quenched. Alternatively, the sample can be forced into contact with a cold metal surface. All the above techniques find application and appear to produce comparable results. Cooling rates 10^4 and 10^5 deg s^{-1} under the most favourable conditions are quoted (Echlin, 1985). The cooling rate can be further enhanced if quenching is carried out under increased pressure, but the technique is complex and subject to various technical problems (Moor & Mühlethaler, 1963).

9.4   Low temperature techniques in electron microscopy

Although the advent of high resolution electron microscopy as applied to labile organic materials has led to remarkable advances in the biological sciences, the technique suffers from actual and potential shortcomings associated with the conditions under which the specimens have to be examined. The combination of high vacuum and high energy electron beams necessitates various pretreatments designed to render the sample resistant to such drastic treatments. Common procedures include chemical fixation, dehydration, embedding in resin, sectioning or fracturing, coating with various substances or replication. At the end of such procedures one must wonder whether the image produced by the electron beam is a true representation of the original, *in vivo* architecture. Where the electron microscope is employed to define the location of low molecular mass, water soluble species, as in X-ray microanalysis, such doubts become certainty, because chemical fixation (e.g. with glutaraldehyde) renders membranes leaky even to macromolecules.

The advantages of freeze fixation methods over the conventional techniques are several: low temperature reduces the rates of diffusive processes and of chemical reactions. It might therefore become possible to monitor kinetic processes by electron microscopy. Low temperature also increases the mechanical strength of materials. Dissection, sectioning and fracturing might thus be possible without the necessity for resin embedding. Further, there is no need for deleterious chemical fixation treatments if the tissue can be fixed in its fully hydrated state. Finally, the incidence of radiation damage is much reduced at very low temperatures of observation.

The disadvantages of low temperature fixation have already been hinted at. Under the most favourable conditions ice crystallization can be completely prevented (vitrification), but in most cases this is hardly a realistic aim. The ice crystal dimensions must be controlled to remain well below the dimensions of the biological features to be examined. At worst, recalescence occurs, leading to massive recrystallization and gross disorganization and distortion of the biological architecture. In order to prevent recrystallization, the specimen must be cooled rapidly and kept below the devitrification temperature of ice, 143 K. It is too often assumed that solidification is complete below 253 K. Not only is this a mistake but even if it were true, such a high temperature would favour recrystallization and coarsening of the texture, and it would also favour the sublimation of ice (freeze drying). The general rule must be to cool as rapidly as possible. Thin sections or cell suspensions in the form of droplets might then be vitrified or partly vitrified. For larger samples the surface layer exposed to the cryogen or cooled metal block might be vitrified to a depth of $< 20\,\mu$m. The general principle must be to achieve maximum undercooling and nucleation and to control the subsequent growth of microcrystals. Just as cyroprotectants are used to control freezing during the low temperature preservation of biological material, so chemicals can be used to facilitate vitrification and to retard crystal growth. Ideally such chemical cryofixing agents should not interfere with physiological function or even render the plasma membrane leaky, especially where the final objective is the microanalysis of soluble cytoplasmic components.

Presumably because of its long-standing use as a cryopreservative, glycerol has in the past been frequently employed to control the ice crystal size in electron microscopic preparations. However, careful comparisons of glycerol treated and untreated specimens suggest that the cryofixative may affect the very features which are to be examined, for instance the distribution of proteins in a plasma membrane (McIntyre, Gilula & Karnovsky, 1974; Böhler, 1975). Any morphological disturbance caused by the cryofixative will be the more serious the higher is the resolution to be achieved. The aim must be to transform labile *in vivo* structures and molecular distributions into a solid state without any disturbance. In the presence of water soluble additives, especially those that can penetrate the plasma membrane, there must always be some doubt whether the features finally observed are not partly the result of some perturbation caused by the additive. In recent years a new class of cryofixatives has achieved some popularity: high molecular mass water soluble polymers, in particular dextran, polyvinyl pyrrolidone (PVP), hydroxyethyl starch (HES) and polyethylene glycols (PEG) (Skaer, Franks, Asquith & Echlin, 1977; Skaer, Franks & Echlin, 1978; Franks, 1980). These substances are completely miscible with water. The viscosities of their aqueous mixtures can be regulated by their concentrations and respective molecular masses. Since they cannot penetrate cell membranes, they exert an osmotic pull on the cell water, but such stress is much less severe than that produced by equivalent mass concentrations of low molecular mass additives, such as sucrose. Their main advantage is that their aqueous solutions can be easily vitrified, so that freezing of the extracellular fluids can be completely inhibited. The plastic properties of aqueous polymer glasses also render them suitable for sectioning at low subzero temperatures, so that they have found application as embedding media.

The vitrification and devitrification of an aqueous polymer solution is graphically illustrated in Fig. 9.5. A rapidly quenched solution of PVP, subjected to fracturing and replication, has the characteristic microspherical appearance shown in Fig. 9.5*a*. There is no indication of ice crystallization, or even nucleation. The microspheres consist of hydrated polymer aggregates. When the quenched preparation is briefly heated to 253 K and then cooled once again to 170 K, fractured and replicated, it takes on the appearance shown in Fig. 9.5*b*. After sublimation (etching), the positions that were occupied by hexagonal ice crystals are now clearly visible, and the residual polymer solution has been freeze concentrated to a composition corresponding to T'_g in Fig. 3.10. For PVP–water mixtures $T'_g = 233$ K at a water content (unfreezable water) of 35% by weight (MacKenzie & Rasmussen, 1972; Franks *et al.*, 1977). Similar microspherical morphologies have been produced from quenched aqueous solutions of HES, PEG and polyvinyl alcohol (Franks, Asquith, Skaer & Roberts, 1979).

Where water is to be removed subsequent to quench freezing, e.g. by sublimation, then there are distinct disadvantages attached to prior vitrification. In the first place the cryofixative is nonvolatile and cannot be removed without an additional solvent treatment. More important, since water is one of the components of the glass, its diffusive motions are

Fig. 9.5. Quench fracture electron micrographs of 50% aqueous PVP solutions. *a*: quenched solution fractured at 123 K, showing complete vitrification and characteristic microspherical morphology; *b*: quenched solution rewarmed to 250 K, cooled to 123 K, fractured and etched. C = ice crystals, VM = vitreous matrix. The scale markers represent 100 nm. Reproduced from Franks *et al.* (1977).

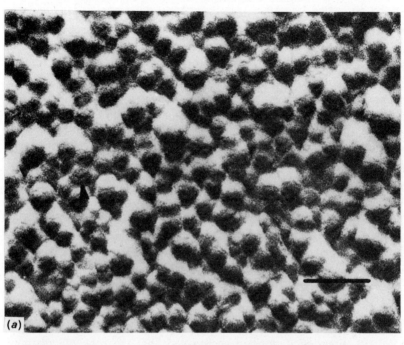

(a)

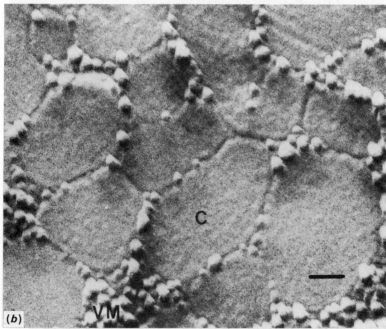

(b)

severely inhibited to the extent that sublimation in a reasonable time becomes impossible in practice.

Despite its attendant practical problems, low temperature microscopy must be the most promising approach to better methods of analysing the ultrastructural and compositional details of labile, water based materials.[†]

9.5   Chilling and freezing in food processing and storage

Although commercially of great value and importance, the freezing of food and its preservation in frozen storage is, from a scientific and technological viewpoint squarely based on history and folklore. One can do no better than to quote from *Recommendations for the Processing and Handling of Frozen Foods*, published by the International Institute of Refrigeration in 1972:

> Food is in the frozen state when a high proportion of the freezable water is present as ice. Frozen food should have been subjected to a freezing process specially designed to preserve the wholesomeness and quality of the product by minimizing physical, biochemical and microbiological changes both in the freezing process and during subsequent storage.
>
> When the term 'Quick frozen' or 'Deep Frozen' is used it generally means food is frozen and preserved in the following manner:
> – Freezing is done in such a way that the zone of maximum crystallization (for most products between −1 and −5 °C) is passed through quickly and the freezing is complete only when the equilibrium temperature reaches −18 °C;
> – The product temperature is maintained at −18 °C or a colder

---

[†] A word of caution is needed on possible connections between high fidelity electron micrographs and low temperature preservation of biological materials, as discussed in Chapter 8. It is the aim of a microscopist to produce 'beautiful' micrographs which exhibit a maximum of structural detail. Such photographs are believed to reflect the 'true' *in vivo* state of the specimen before its long journey began which ended on the metal grid inside the electron microscope.

The cryobiologist is concerned with devising treatments which will enhance the probability of survival of living tissue after prolonged periods of storage at subfreezing temperatures. We have seen that osmotic dehydration and freeze concentration figure largely among common treatments, and the appearance of the tissue in its artificially cold hardened state is marked by cell shrinkage and distortion. It is in fact the very opposite in appearance from the beautiful electron micrographs which are so common in the biological literature (Farrant, 1977). What is certain, despite claims to the contrary (Moor & Mühlethaler, 1963), is that the pretreatments commonly used in electron microscopy to maintain a beautiful true-to-life (?) *appearance* are the very opposite of those used by the cryobiologist to promote preservation of *function*.

temperature during storage and transport with a minimum temperature variation.

Quite apart from semantic problems with expressions like 'wholesomeness' and 'quality', the above definitions beg several questions. For instance, it is by no means the case that no residual liquid phase exists below $-18\,°C$. Indeed, the choice of this particular temperature for deep frozen storage may well be responsible for loss of quality and wholesomeness, however defined. It is also by no means proven that the internationally recommended protocols for freezing and storage do in fact minimize physical and chemical changes. There is much evidence to the contrary. The deleterious effects of freeze concentration are ever present and bad management of ice crystallization produces inferior textures. One of the most undesirable manifestations of such freeze damage is 'drip loss', the inability of the thawed product to hold water in its tissues. Altogether it is surprising that an industry that prides itself on its scientific and technological sophistication is largely unaware of the basic physical and chemical principles, described in Chapters 2 and 3, which govern the behaviour of aqueous systems at subzero temperatures.

There can be no doubt that the successful low temperature processing and storage of commodity foodstuffs is beset by severe technical problems. A typical example of such problems is provided by the handling of meat. Post-mortem anaerobic glycolysis in animal muscle cannot produce enough ATP to prevent substantial cross-linking of the muscle actomyosin. The resulting loss of elasticity is known as *rigor mortis*. The glycolysis produces lactic acid, causing a fall in the pH of the tissue from its *in vivo* value of 7.2 to 5.5 which also happens to be close to the isoelectric point of the muscle proteins, corresponding to a minimum hydration capacity and extensive drip loss on thawing. The quantity of glycogen present in the muscle at the time of slaughter will determine the final pH, and the glycogen content itself depends on the degree of pre-slaughter stress to which the animal has been subjected. Although a higher pH provides for better water holding, it is generally undesirable because it also leads to discoloration and an unacceptable texture.

The kinetics of post-mortem glycolysis also affect the attributes of the final product. Thus, by chilling the carcass immediately after slaughter, the rates of pH decrease, ATP depletion and *rigor mortis* onset can be retarded. In the case of beef and lamb muscle, however, the rate of ATP hydrolysis actually begins to increase as the temperature is lowered to $< 15\,°C$, leading to enhanced toughness of the meat.

It might be thought that rapid freezing would arrest post-mortem glycolysis, maintain a high pH level and prevent the breakdown of ATP.

This may well be so, but all three damaging processes then take place even more rapidly during thawing (thaw *rigor*). If meat which has been frozen pre-*rigor* is prevented from contracting during thawing, e.g. by keeping it on the bone, the deleterious effects of freezing can be reduced. Alternatively, the meat can be stored in the frozen state for some months, during which period the ATP will be decomposed, so that thaw *rigor* is minimized.

Even after the termination of anaerobic glycolysis and lactic acid production, other changes take place in muscle which affect its response to freezing. Some of them are direct results of proteolytic enzyme activity and others are of chemical origin, e.g. fat oxidation. Proteolytic activity can be beneficial because of its tenderising effect. For this reason beef is frequently conditioned at 0 °C for periods of up to two weeks.

Apart from the chemical and biochemical changes associated with meat preservation, attention must also be focussed on the physical and engineering aspects of freezing (Calvelo, 1981). Thus, meat being a cellular material, freezing is generally initiated in the extracellular fluids. Intracellular nucleation can only be achieved at very high cooling rates. In meat technology a characteristic time t_c is defined as the time required for the temperature at any given point to fall from -1 to -7 °C, corresponding to the freezing of 80% of the water present. The condition for intracellular

Fig. 9.6. Ice crystal dimensions in frozen beef as a function of t_c, the time required for the temperature to decrease from -1 to -7 °C. Intracellular (●) and extracellular (○) ice crystals depend on t_c. After Calvelo (1981).

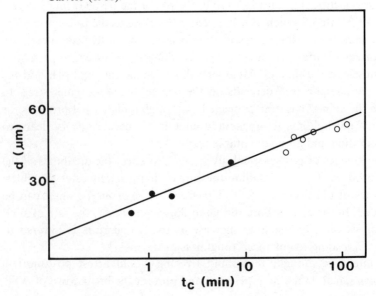

nucleation is that $t_c < 5$ minutes, which, in beef, can only be achieved for a depth of < 2 cm (refer to Fig. 2.7.). Figure 9.6 serves to illustrate this point. For a ΔT of 4 °C and a linear temperature distribution, dendrites grow at a rate of 0.2 cm s^{-1}. When the free dendrite domain is consolidated $(T = T_f)$, growth of cellular dendrites begins and large, thick crystals are formed. For reasons which are not clear, intracellular crystals always grow longitudinally to the fibre axis but have a limited growth with $t_c = 20$ min. At a greater depth within the tissue crystallization proceeds by dehydration of the meat fibres. Experience shows that intracellular freezing is limited to a depth of 1–2 cm, so that most of the ice formed in meat originates in the extracellular domains and its growth is accompanied by osmotic dehydration of the cells and gross distortion of the tissue. The temperature profile in the tissue depends on the initial temperature, the temperature of the coolant, thermal conductivities and thermal transfer coefficients and the density, thickness and specific heat of the material. For calculations of practical value a model is required for the solution of eqn (9.6) which can take into account the unidirectional freezing of a heterogeneous

Fig. 9.7. Temperature profile during the asymmetric freezing of beef (heat flux perpendicular to the fibre axis). The total thickness of the sample is 2L. After Calvelo (1981).

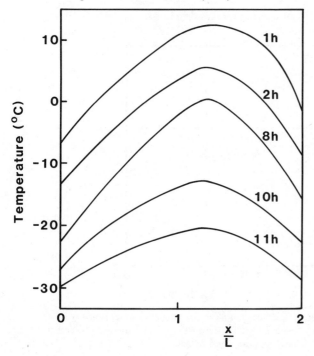

material; generally the temperature gradient is normal to the direction of the meat fibres. In engineering calculations the problem is simplified: the time required for complete freezing is estimated (i.e. the time taken to reach $-18\,°C$). In reality this is not a very satisfactory assumption, since even at $-60\,°C$ there is still a measurable amount of unfrozen, mobile water present. Figure 9.7 shows the calculated temperature profile for the asymmetric freezing of beef; the heat flux is perpendicular to the meat fibre axis.

The efficiency of a particular freezing protocol is judged by the amount of drip loss produced after thawing; this is shown in Fig. 9.8. Maximum water loss correlates well with the incidence of intracellular freezing which produces single crystals inside the meat fibres. On the other hand, extracellular freezing at the expense of cell dehydration is independent of the cooling rate and only depends on the temperature. Cell dehydration is also seen to be partly reversible because some water is reabsorbed by the cells during thawing, indicating that the cell membrane can still function as an osmometer.[†] Thawing is thus seen to be another factor in determining quality. It is not as easily controlled as freezing because it is limited by the thermal conductivity of ice and also because thawing is performed by the consumer in the home under uncontrolled conditions.

Fig. 9.8. Effect of freezing rate on drip loss in beef. After Calvelo (1981).

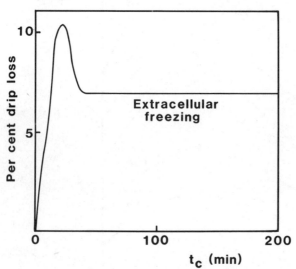

† It must be noted, however, that a certain amount of water is in any case reabsorbed into the muscle fibres during thawing, irrespective of whether the cell membranes have been destroyed or not.

Finally consideration must be given to changes that take place during the storage of meat in the freezer. These result mainly from freeze denaturation of the proteins, the sublimation of ice (freezer burn) and recrystallization/ripening. The changes in the ice crystal size distribution in beef are shown in Fig. 9.9. Grain growth takes place at all temperatures but is accelerated when small crystals, grown at low temperatures (optimum freezing conditions) are subjected to higher storage temperatures. From the observed changes in the crystal size distribution it is possible to estimate that 60 μm diameter crystals double in size after 21 days storage at $-10\,°C$ or 50 days storage at $-20\,°C$.

The above summary of the chief factors governing the freezing of meat emphasizes the complex interplay of a variety of physical and biochemical factors. The optimization of the process would probably be aided by a better understanding of these factors. The situation is even more complicated in the case of fish processing. Because of the lower glycogen content of fish muscle, the post-mortem fall in pH is not very marked. The tissue is therefore more susceptible to bacterial growth and spoilage. In addition the high proportion of polyunsaturated lipids makes fish muscle tissue prone to autoxidation during frozen storage. The proteins of fish also differ from those in meat in that they suffer freeze denaturation, particularly in the high subzero temperature range.

Fig. 9.9. Effects of frozen storage on ice crystal dimensions in beef. For details see text. After Calvelo (1981).

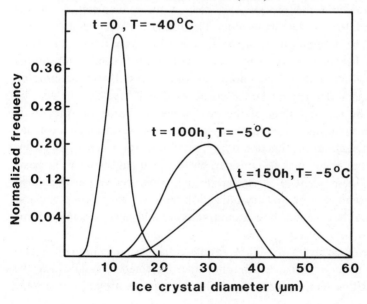

Whereas commodity foods are frozen mainly to prolong their shelf life, other types of products derive their benefits from being eaten cold. Ice cream will serve as an example for a discussion of the various processing variables (Berger, Bullimore, White & Wright, 1972). The product itself is a multiphase disperse system which contains 50–55% by volume of air. The weight composition of the mix is 6–12% fat, 7–12% milk protein solids, 13–18% sugar, 64% water (mainly as ice) and minor amounts of emulsifiers and stabilizers. The manufacturing sequence is as follows: mixing, pasteurization, homogenization, cooling, ageing, aeration and freezing, extrusion at $-6\,^{\circ}$C, packaging, hardening at $-35\,^{\circ}$C and storage/distribution at $-11\,^{\circ}$C. The eating quality of the finished product is determined by the distributions of air cells, ice crystals and fat globules within the matrix and by the rheological properties of the various interfaces: fat–aqueous phase, air–aqueous phase, ice–aqueous phase and by the various interfacial films where the protein is concentrated. Major changes take place in the aqueous phase during processing. The percentage of frozen water increases from 50% during extrusion to 95% during hardening and then falls to 72% during storage. During the hardening stage the residual aqueous phase (excluding ice) is a concentrated (86%) sugar glass containing sucrose and lactose; it has a $T'_g = -23\,^{\circ}$C. The phenomenon known as 'grittiness' is due to the crystallization of lactose which is detectable at $> -23\,^{\circ}$C and becomes rapid at $> -17\,^{\circ}$C. It is prevented by reducing the milk serum solids.

Considering the ice crystals first, their size distribution is affected by the composition and the process variables; a mean diameter of 40 μm corresponds to a satisfactory product. The distance between crystals is then 6–8 μm. The destabilization of the fat emulsion gives rise to larger ice crystals, whereas an increase in the amount of water soluble substances depresses T_f and reduces the amount of water. Crystals will therefore be smaller and separated by larger distances. Small crystals can also be achieved by rapid cooling of the mix, while the hardening process is accompanied by an increase in the crystal dimensions. Although much has been written about the function of polysaccharide stabilizers in ice cream manufacture, there must be considerable doubt about the scientific basis of most of these reports. Polysaccharide gums have been variously claimed to affect ice nucleation and crystallization rates, crystal morphology and maturation, but the evidence is contradictory and unconvincing (Muhr, 1983).

Storage of ice cream at $-11\,^{\circ}$C results in the partial melting of ice and a shift in the crystal size distribution, since the smaller crystals have the lowest melting temperatures. Similarly, fluctuations around the storage temperature produce a gradual coarsening of the texture, because on cooling, water condenses on existing crystals in preference to forming new ones. Once the ice crystal diameter exceeds 55 μm, ice cream becomes organoleptically unacceptable. The shape, size and distribution of ice crystals in the aqueous phase are thus seen to be sensitive to the particular manufacturing process and the storage regime, but these parameters also contribute in a significant manner to the perceived eating quality of the product. Other factors of equal importance in determining the texture are the fat emulsion and the nature of the milk protein film at the water–oil and water–air interfaces. The rheology of the interfacial films depends on the ion (Ca^{2+}) concentration, on pH and on the details of the heat treatment to which the mix is subjected during pasteurization. On the other hand, it is thought that the interfacial properties of air–mix and ice–mix interfaces are of less importance, because the number ratio of air bubbles:ice crystals:fat globules is $1:1:2 \times 10^5$.

The technology of ice cream production, storage and distribution is seen to be complex, but a close examination of the various steps in interactions demonstrates that several of the principles described in earlier chapters in this book are in fact implicated. There does not appear to exist a thorough-going analysis of the manufacturing processes employed in terms of these basic physical principles.[†] These principles also form the basis of other, more recent developments in frozen and chilled food product technology. For instance, the Rich Company of Buffalo, USA, has patented a range of products (Freeze-Flo) which can be stored in a domestic freezer without becoming hard and brittle. In addition to their obvious convenience (no thawing required), such products are organoleptic curiosities. Even when eaten 'straight from the freezer' they hardly produced a cold sensation, because with their very low ice content, very little latent heat is absorbed in the mouth.

The products contain high concentrations of additives of carbohydrate origin which act as freezing point depressants. The company has described the products as resulting from the application of 'tomorrow's technology'. It is amusing to contemplate that insects have been relying on these same processes as a means of survival for millions of years. It seems, however, that the exploitation of these techniques by insects, plants and micro-organisms does not constitute 'prior art' within the terms of patent law.

[†] It is possible, although unlikely, that such rigorous scientific investigations have been conducted in the research and development departments of one or several of the major food manufacturing companies but that the results were considered to be classified information, to be hidden away in the archives rather than disseminated.

10

Matters arising

By way of drawing together the various aspects of the impact of low temperatures on the physics and chemistry of life processes, it is helpful to identify the areas where hard fact and plausible theory merge into speculation and folklore. There can be no doubt that the energetics and kinetics of life processes depend intimately on the physical properties of the aqueous substrate, which in turn exhibits a marked temperature sensitivity, especially at the lower end of the physiological temperature range. Life on earth is subject to a number of physiological stress conditions; of these cold, salinity and drought, in that order, are the most important and widespread. Further stress is provided by temperature fluctuations. In the Himalayas their amplitudes may exceed 50 degrees in a 24 hour period. Any mechanism of adaptation must then be geared to cope with such fluctuations. In this book I have emphasized the effects of low temperature on the physical properties of water and aqueous solutions. The results of such investigations have then been related to the subtle biochemical and biophysical changes in biomacromolecules which form the basis of life. At the next level of complexity, physiological and biochemical consequences of chill and freezing tolerance have been discussed in terms of the physico- and biochemical principles earlier established. The exploitation of these same principles in the preservation of live tissue and the technology of low temperature processing of labile materials forms the subject matter of the concluding chapters.

It is evident that progress in our understanding of the many interrelated processes is beset by large gaps in our knowledge, some of them vital, others of lesser significance, but still of interest. It is the purpose of this concluding chapter to identify such areas of uncertainty and to indicate possible lines which purposeful investigations might take.

10.1   Water and aqueous solutions

The physical properties of undercooled water are of most immediate interest. Thanks mainly to Angell and his associates (Angell, 1982,
1983) we now possess a wealth of reliable experimental data. The most
striking feature which most of these properties have in common is their
apparent divergence at 228 K. So far there is no satisfactory explanation
for the observations. The divergence may be an indication of a critical type
phenomenon or a physical instability (spinodal) in the liquid state, i.e. a
discontinuity in the free energy surface. Closely related to the divergence
is the observation that more normal behaviour is achieved by the addition
of solutes, implying that the particular features that make water unique
are destroyed by moderate concentrations of salts, hydrogen peroxide or
hydrazine (Angell, 1982).

Another unresolved puzzle concerns the relationship between undercooled liquid water and amorphous ice, produced by the deposition of
water vapour on cold (< 120 K) surfaces. Views have been advanced
favouring a continuity of state between amorphous ice and liquid water
(Sceats & Rice, 1982; Mayer & Brüggeller, 1983); on the other hand the
existence of the singularity at 228 K, referred to above, is hardly compatible
with such claims (Angell, 1982). Closely related, and of much more
immediate practical importance, is the question of whether liquid water
can be rapidly quenched into the glassy state and whether such a state can
indeed be identical to amorphous ice, bearing in mind that liquid water
is extensively hydrogen bonded, whereas vapour-deposited amorphous ice
is not.

The theory of nucleation of crystalline phases from the melt is based on
a model of stepwise growth of a cluster until its dimensions become such
that further condensation is spontaneous. On the whole, this model has
been able to account reasonably well for the nucleation of metals, salts and
simple molecular crystals in the melt or in solution. The case of water
appears to be unique. Indications are that the density fluctuations (on
which nucleation depends) are highly cooperative (Bosio, Teixeira &
Stanley, 1981) and become increasingly so as 228 K is approached. It is
questionable, therefore, how the stepwise growth model can adequately
account for the nucleation of ice in undercooled water. A recent re-
examination of the nucleation process suggests that it cannot (Franks,
Mathias & Trafford, 1984). Following on from recent studies of solutions
of sugar and sugar alcohols, indicating the existence of interesting stereo-
specific effects on the hydration properties of such molecules (Franks,
1979) and their interactions with one another (Barone, Bove, Castronuovo
& Elia, 1981), it is to be hoped that more comprehensive investigations

into the solution behaviour will receive fresh impetus. Their involvement in physiological processes such as freeze avoidance and freeze tolerance has been amply demonstrated, but a molecular description of their solutions is still wanting. Closely related is the observation that such compounds are able to protect proteins against dehydration, whether by drying or freezing. Although the thermodynamic parameters associated with such effects have been established, there is as yet no explanation for the apparent ability of sugars and sugar alcohols to 'salt-out' proteins (Gekko & Morikawa, 1981).

Concentrated solutions of sugars and other carbohydrate derivatives have not yet received the attention they clearly deserve in view of their ecological and technological importance. Solid–liquid state diagrams of the type shown in Fig. 3.10 are urgently required, as is also better information about temperature–concentration–viscosity relationships, so that reliable extrapolations can be made to estimate glass transition temperatures and unfreezable water contents. It is probably unrealistic to expect a unified theory for such mixtures to emerge in the forseeable future. From the point of view of technological applications of subfreezing temperatures, more and better data are also required on molecular diffusion and on the kinetics of reactions in viscous solutions. At the present time little is known about the rates and mechanisms of reactions in concentrated systems containing also chemically inert substances such as salts or sugars.

Somewhat related are the problems surrounding the glassy aqueous states referred to in Chapter 9, and illustrated in Fig. 9.5, especially where the nonaqueous component is a macromolecule. In such a system, where water fulfils the role of a low molecular mass plasticizer, which is, however, liable to crystallization, a perceptive study of the diffusional motions should be rewarding. It is well known that water molecules in an aqueous gel suffer very little inhibition of their motional freedom, despite the fact that the system as a whole has the mechanical properties of a solid. It is not inconceivable that at a given temperature the diffusion rates of the various components in an amorphous solid might differ by orders of magnitude. This raises questions about the significance of the measured glass temperature of such mixtures and their long term stabilities.

10.2   Cryobiochemistry

The problems of protein folding and the stability of folded states are currently receiving much attention. Diffraction (by crystals) coupled with computer simulation (*in vacuo*) and spectroscopy (in solution) are said to provide a powerful combination with which to probe the forces responsible for the maintenance of biologically active protein conformations. It is symptomatic of the state of the art that in an almost 600 page volume devoted to this subject no reference is made to the solvent medium or its possible function in the generation and stabilization of such folded structures (Jaenicke, 1980). While lip service is generally paid to the role of solvation, few attempts are made to quantify the solvent contribution to the observed phenomena. It is obvious that the generalized free energy surface shown in Fig. 4.4 is dominated by solvent effects, specifically by the large partial heat capacities of the peptide chains in water. It is difficult to imagine how computer studies of protein dynamics, based on isolated peptide chains *in vacuo*, will ever explain the marginal and temperature sensitive stabilities of proteins *in vitro*, let alone *in vivo*. At the same time, it is exactly the marginal nature of the stability which is responsible for biological function.

Granted that there is not at present a satisfactory theoretical method for coping with hydration or hydrophobic interactions, it nevertheless seems irrelevant to perform complicated and costly computer calculations on isolated peptide chains. The real problems are being side-stepped in this way. One can only hope that the central role played by water in modulating *in vivo* processes will eventually be recognized by the protein theoreticians. As pointed out earlier, a study of the *low* temperature stability limit should pay particular dividends. For an order/disorder transition to occur upon a reduction of kinetic energy is rare, possibly unique to aqueous systems in which hydrophobic and electrostatic effects dominate the observed behaviour.

The practical possibility of protein experimentation in undercooled water (Douzou, Debey & Franks, 1978; Douzou, Balny & Franks, 1978) should provide stimulus to experimental investigations of cold lability in proteins. The technique may also partly replace the conventional method of cryoenzymology, involving the use of mixed organic/aqueous solvents (Douzou, 1977). Low temperature protein studies are of particular interest in connection with the biochemistry of cryoprotectant production in living organisms during cold hardening.

10.3 Mechanisms of chill and frost hardening

Much remains to be done to elucidate the biophysical and biochemical details of seasonal protection against cold injury, whether caused by chill or by freezing. Of the two, chill hardening is probably easier to understand because it does not involve osmotic gradients across the cell membrane. Apart from a vague appreciation that both temperature and photoperiod are involved in the triggering of the hardening process, we know little of the mechanistic details of such triggers, or of the processes which set in motion dehardening.

In vivo nucleation of ice, or its inhibition, is another general area of ignorance. It is by no means obvious how peptide chains, apparently without any regular structures, can interfere with the nucleation and/or growth of ice, while other peptides, presumably with a highly regular structure, are able to reduce undercooling to almost zero by efficient ice nucleation (Duman, 1982). Models which have in the past been advanced, based largely on stereocompatibility between water or ice and the antifreeze peptide, must be considered as simplistic and unconvincing (Franks & Morris, 1978: De Vries, 1983). Perhaps the most mysterious effect in *in vivo* cold resistance is that of deep undercooling, where large volumes of water appear to be prevented from freezing at temperatures approaching, or even below, the normal homogeneous nucleation temperature for very small volumes, $-40\,°C$. The possibility that hardening involves the segregation of the tissue fluids in small volumes cannot be ruled out, and we must await the results of detailed histological and ultrastructural studies on hardened tissues.

The fact that hardening (whether against cold, heat or saline conditions) is often accompanied by the biosynthesis of appreciable concentrations of various water soluble chemicals within the cells is a sure indication that survival is not directly determined by the water activity, since it seems to be exactly the achievement of *lower* a_w values that produce the internal environment necessary for survival.

10.4   Cryobiology and cryotechnology

The point was made in Chapter 8 that cryobiology, the science of low temperature preservation of live tissues, is still very much a recipe science. A much better understanding of the basic processes involved, and their interplay, is required before the routine cryopreservation of organs can become a feasible proposition. At the present time even the preservation of many types of single cells is still impossible, for reasons unknown.

Whether the injurious effects of freeze concentration can be overcome remains to be seen, but a better understanding and quantification of the mass fluxes and concentration transients that take place in a tissue during the application of a temperature gradient is imperative. The modelling of cells and tissues as 'bags filled with water' and surrounded by semipermeable membranes may be an encouraging beginning (Silvares *et al.*, 1975; Diller & Lynch, 1983), but such simple models cannot be expected to account adequately for the real response of the tissue to cooling and freezing, especially in the presence of cryoprotective additives. The direct observation of the physical processes involved by means of the cryomicroscope may provide results against which model predictions can be tested. Apart from the quantity of ice produced which determines the degree of freeze concentration, the rate of freezing and the ice crystal morphology may also be determinants of injury and preservation in the case of tissues and whole organs where the extracellular matrix is subject to mechanical damage by ice crystals. Little is yet known about the control of freezing in complex heterogeneous substrates, especially those of biological origin.

Most of the technological processes that rely on freezing or part freezing are firmly based on folklore. Too frequently experimental observations on highly complex systems, and based on measurements performed under nonequilibrium conditions, are rationalized in terms of elementary textbook science.[†]

The involvement and importance of metastable water in many industrial processes can hardly be overestimated (Franks, 1984). Most of the technical problems are associated with the exercise of proper kinetic control of ice crystallization during cooling. A fundamental study of the kinetics is beset by difficulties associated with the chemical complexities of many of the systems of interest, e.g. ice cream, and with the very high viscosities. Nevertheless, the properties and usefulness of the final product, whether freeze-thawed, freeze-dried or used in the frozen state, depend on the degree of undercooling and supersaturation, the mechanism of the nucleation of ice and other constituents, the growth and type of crystals, their size distribution, the flow properties of the concentrated, unfrozen matrix and any long term changes due to ageing. Fundamental studies of such processes are unglamorous, tedious and time-consuming, but they pay dividends in terms of quality control and flexibility in product and process design. They certainly form the scientific base of low temperature technology, the understanding and exploitation of which is still in its infancy, even among those who should know better.

[†] A classic example of the absurd conclusions to which such simplistic reasoning can lead is provided by an analysis of cryoscopic data on concentrated aqueous solutions (Ross, 1954). The author himself comments on 'a serious theoretical discrepancy' between the predicted and actual cryoscopic behaviour of aqueous solutions of various diols. However, he goes on to state that 'in spite of this discrepancy, this [his theory] concept was found to be of practical value...'

References

Angell, C. A. (1982). Supercooled water. In *Water – A Comprehensive Treatise*, vol. 7, ed. F. Franks, pp. 1–82. New York: Plenum Press.

Angell, C. A. (1983). Supercooled water. *Ann. Rev. Phys. Chem.*, **34**, 593–630.

Arakawa, T. & Timasheff, S. N. (1982). Stabilization of protein structure by sugars. *Biochem.*, **21**, 6536–44.

Ashwood-Smith, M. J. (1980). Low temperature preservation of cells, tissues and organs. In *Low Temperature Preservation in Medicine and Biology*, ed. M. J. Ashwood-Smith & J. Farrant, pp. 19–44. Tunbridge Wells: Pitman Medical Ltd.

Ashwood-Smith, M. J. & Farrant, J. F. (1980). *Low Temperature Preservation in Medicine and Biology*. Tunbridge Wells: Pitman Medical Ltd.

Avrami, M. (1941). Granulation, phase change and microstructure, III. Kinetics of phase change. *J. Chem. Phys.*, **9**, 177–84.

Bagnall, D. J. & Wolfe, J. (1982). Arrhenius plots: information or noise? *Cryo-Letters*, **3**, 7–16.

Bajaj, Y. P. S. (1976). Regeneration of plants from cell suspensions frozen at −20, −70 and −196 °C. *Physiol. Plant.*, **37**, 263–8.

Barone, G., Bove, B., Castronuovo, G. & Elia, V. (1981). Excess enthalpies of aqueous solutions of polyols at 25 °C. *J. Solution Chem.*, **10**, 803–9.

Barthel, J., Gores, H. J., Schmeer, G. & Wachter, R. (1984). Non-aqueous electrolyte solutions in Chemistry and modern technology. In *Topics in Current Chemistry*, **111**, 33–144.

Baust, J. G., Lee, R. E. & Ring, R. A. (1982). The physiology and biochemistry of low temperature tolerance in insects and other terrestrial arthropods: a bibliography. *Cryo-Letters*, **3**, 191–212.

Berger, K. G., Bullimore, B. K., White, G. W. & Wright, W. B. (1972). The structure of ice cream. *Dairy Industries*, **37**, 419–25.

Bigelow, C. C. (1967). On the average hydrophobicity of proteins and the relation between it and protein structure. *J. Theor. Biol.*, **16**, 187–211.

Bigg, E. K. (1953). The supercooling of water. *Proc. phys. Soc.* **B66**, 688–703.

Biswas, A. B., Kumsah, C. A., Pass, G. & Phillips, G. O. (1975). The effect of carbohydrates on the heat of fusion of water. *J. Solution Chem.*, **4**, 581–90.

Blackburn, G. M., Lilley, T. H. & Walmsley, E. (1980). Aqueous solutions containing amino acids and peptides. *J. Chem. Soc. Faraday Trans.* 1, **76**, 915–22.

Bock, P. E. & Frieden, C. (1976). Phosphofructokinase. I. Mechanism of the pH dependent inactivation and reactivation of the rabbit muscle enzyme. *J. Biol. Chem.*, **251**, 5630–43.

Bock, P. E. & Frieden, C. (1978). Another look at the cold lability of enzymes. *Trends in the Biochemical Sciences*, May 1978, 100–3.

Böhler, S. (1975). *Artefacts and Specimen Preparation Faults in Freeze Etch Technology*. Liechtenstein: Balzers AG.

Bosio, L., Teixeira, J. & Stanley, H. E. (1981). Enhanced density fluctuations in supercooled H_2O, D_2O and ethanol–water solutions: evidence from small-angle X-ray scattering. *Phys. Rev. Lett.*, **46**, 597–600.

Brandts, J. F. (1964). The thermodynamics of protein denaturation. I. *J. Amer. Chem. Soc.*, **86**, 4291–301.

Brandts, J. G., Fu, J. & Nordin, J. H. (1970). The low temperature denaturation of chymotrypsinogen in aqueous solution and in frozen aqueous solution. In *The Frozen Cell*, ed. G. E. W. Wolstenholme & M. O'Connor, pp. 189–208. London: J. & A. Churchill.

Brandts, J. F. & Hunt, L. (1967). The thermodynamics of protein denaturation. III. *J. Amer. Chem. Soc.*, **89**, 4826–38.

Brüggeller, P. & Mayer, E. (1980). Complete vitrification of pure liquid water and dilute aqueous solutions. *Nature*, **288**, 569–71.

Buchanan, E. E. & Fulmer, E. J. (1930). *Physiology and Biochemistry of Bacteria*. London: Bailliere, Tindall & Cox.

Burchard, W. (1983). Solution thermodynamics of nonionic water soluble polymers. In *Chemistry and Technology of Water-Soluble Polymers*, ed. C. A. Finch, pp. 125–42. New York: Plenum Press.

Burke, M. J., Gusta, L. V., Quamme, H. A., Weiser, C. J. & Li, P. H. (1976). Freezing and injury in plants. *A. Rev. Pl. Physiol.*, **27**, 507–28.

Calvelo, A. (1981). Recent studies on meat freezing. In *Development in Meat Science*, vol. 2, ed. R. Lawrie, pp. 125–58. London: Applied Science Publishers.

Chapman, D., Williams, R. M. & Ladbroke, B. D. (1967). Physical studies of phospholipids. VI. Thermotropic and lyotropic mesomorphism of some 1,2-diacylphosphatidylcholines (lecithins). *Chem. Phys. Lipids*, **1**, 445–75.

Chen, H. W. & Li, P. H. (1982). Potato cold acclimation. In *Plant Cold Hardiness and Freezing Stress*, vol. 2, ed. P. H. Li & A. Sakai, pp. 5–22. New York: Academic Press.

Chothia, C. (1974). Hydrophobic bonding and accessible surface area in proteins. *Nature*, **248**, 338–9.

Conway, B. E. (1981). *Ionic Hydration in Chemistry and Biophysics*. Amsterdam: Elsevier.

Costello, M. J. (1980). Ultra-rapid freezing of thin biological samples. *Scanning Electron Microscopy*, **11**, 361–70.

Cox, R. P. (1975). Oxygen evolution at subzero temperatures by chloroplasts suspended in fluid media. *FEBS Letters*, **57**, 117–19.

Critchley, C. (1976). 'Untersuchungen an Chloroplastenmembranen über mögliche Veränderungen in den Lipiden im Zusammenhang mit Problemen der Gefrierinaktivierung bzw. der Frostresistenz von Pflanzen.' Unpublished Ph.D. Thesis, University of Düsseldorf.

Crowe, J. H. & Crowe, L. M. (1981). Membrane changes during drying. *Cryobiology*, **18**, 613.

David, P. (1983). Suit filed against NIH. *Nature*, **305**, 262.

Dean, W. L. & Tanford, C. (1978). Properties of a delipidated, detergent-activated Ca ATPase. *Biochem.*, **17**, 1683–90.

Derbyshire, W. (1982). The dynamics of water in heterogeneous systems with emphasis on subzero temperatures. In *Water – A Comprehensive Treatise*, vol. 7, ed. F. Franks, pp. 339–430. New York: Plenum Press.

DeVries, A. L. (1983). Antifreeze peptides and glycopeptides in cold-water fishes. *A. Rev. Physiol.*, **45**, 245–60.

DeVries, A. L., Komatsu, S. K. & Feeny, R. E. (1970). Chemical and physical properties of freezing point-depressing glycoproteins from antarctic fishes. *J. biol. Chem.*, **245** 2901–13.

DeVries, A. L. & Wohlschlag, D. E. (1969). Freezing resistance in some Antarctic fishes. *Science, N.Y.*, **163**, 1074–5.

Diller, K. R. & Lynch, M. E. (1983). An irreversible thermodynamic analysis of cell freezing in the presence of membrane permeable additives. I. Numerical model and transient cell volume data. *Cryo-Letters*, **4**, 295–308.

Diller, K. R. & Lynch, M. E. (1984*a*). *Art. cit.* II. Transient electrolyte and additive concentrations. *Cryo-Letters*, **5**, 117–30.

Diller, K. R. & Lynch, M. E. (1984*b*). *Art. cit.* III. Transient water and additive fluxes. *Cryo-Letters*, **5**, 131–44.

DiPaola, G. & Belleau, B. (1977). Polyol–water interactions. Apparent molal heat capacities and volumes of aqueous polyol solutions. *Can. J. Chem.*, **55**, 3825–30.

Dixon, W. L., Franks, F. & ap Rees, T. (1981). Cold liability of phosphofructokinase from potato tubers. *Phytochemistry*, **20**, 969–72.

Dore, J. C. (1981). Discussion comment. In *Biophysics of Water*, ed. F. Franks & S. F. Mathias, pp. 356–8. Chichester: John Wiley & Sons.

Douzou, P. (1977). *Cryobiochemistry*. New York: Academic Press.

Douzou, P. & Balny, C. (1974). Eventual role of respiratory proteins in cold resistance of certain invertebrates. *Compt. rend. Acad. Sci.*, **279**, 851–3.

Douzou, P., Balny, C. & Franks, F. (1978). New trends in cryoenzymology. I. Supercooled aqueous solutions. *Biochimie*, **60**, 151–8.

Douzou, P., Debey, P. & Franks, F., (1978). Supercooled water as medium for enzyme reactions at subzero temperatures. *Biochim. biophys. Acta*, **523**, 1–8.

Dubochet, J., Adrian, M. & Vogel, R. H. (1983). Amorphous solid water obtained by vapour condensation or by liquid cooling: a comparison in the electron microscope. *Cryo-letters*, **4**, 233–40.

Dufour, L. & Defay, R. (1963). *Thermodynamics of Clouds*. New York: Academic Press.

Duman, J. (1982). Insect antifreezes and ice-nucleating agents. *Cryobiology*, **19**, 613–17.

Duman, J. G. & Patterson, J. L. (1978). The role of ice nucleators in the frost tolerance of overwintering queens in the bald faced hornet. *Comp. Biochem. Physiology*, **A59**, 69–72.

Eagland, D. (1975). Nucleic acids, peptides and proteins. In *Water – A Comprehensive Treatise*, vol. 4, ed. F. Franks, pp. 305–518. New York: Plenum Press.

Echlin, P. (in press). Cryomicroscopy. In *Advanced Techniques in SEM and X-ray Microanalysis*. New York: Plenum Press.

Echlin, P., Skaer, H. leB., Gardiner, B. O. C., Franks, F. & Asquith, M. H. (1977). Polymeric cryoprotectants in the preservation of biological ultrastructure. II. Physiological effects. *J. Microsc.*, **110**, 239–55.

Eisenberg, H. (1976). *Biological Macromolecules and Polyelectrolytes in Solution*. Oxford: Clarendon Press.

Ellory, J. C. & Willis, J. S. (1981). Phasing out the sodium pump. In *Effects of Low Temperatures on Biological Membranes*, ed. G. J. Morris & A. Clarke, pp. 107–20. London: Academic Press.

Enderby, J. E. & Nielson, G. W. (1979). X-ray and neutron scattering by aqueous solutions of electrolytes. In *Water – A Comprehensive Treatise*, vol. 6, ed. F. Franks, pp. 1–45. New York: Plenum Press.

Engberts, J. B. F. N. (1979). Mixed aqueous solvent effects on kinetics and mechanisms of organic reactions. In *Water – A Comprehensive Treatise*, vol. 6, ed. F. Franks, pp. 139–238. New York: Plenum Press.

Fahy, G. M., MacFarlane, D. R., Angell, C. A. & Meryman, H. T. (1984). Vitrification as an approach to cryopreservation. *Cryobiology*, **21**, 407–26.

Farrant, J. (1966). The preservation of living cells, tissues and organs at low temperatures: some underlying principles. *Lab. Practice*, **15**, 402–4.

Farrant, J. (1977). Water transport and cell survival in cryobiological procedures. *Phil. Trans. R. Soc. Ser. B*, **278**, 191–205.

Farrant, J. (1980). General observations on cell preservation. In *Low Temperature Preservation in Medicine and Biology*, ed. M. J. Ashwood-Smith & J. Farrant, pp. 1–18. Tunbridge Wells: Pitman Medical Ltd.

Fennema, O. (1975). Reaction kinetics in partially frozen aqueous solutions. In *Water Relations of Foods*, ed. R. B. Duckworth, pp. 539–58. London: Academic Press.

Finney, J. L. (1979). The organization and function of water in protein crystals. In *Water – A Comprehensive Treatise*, vol. 7, ed. F. Franks, pp. 47–122. New York: Plenum Press.

Finney, J. L. (1982). Solvent effects in biomolecular processes. In *Biophysics of Water*, ed. F. Franks & S. F. Mathias, pp. 55–8. Chichester: John Wiley & Sons.

Fishbein, W. N. & Winkert, J. W. (1979). Parameters of freezing damage to enzymes. In *Proteins at Low Temperatures*, ed. O. Fennema, pp. 55–82. *Adv. Chem. Ser. No. 180.* American Chemical Society.

Fletcher, N. H. (1970). *The Chemical Physics of Ice*. Cambridge University Press.

Frank, H. S. & Wen, W. Y. (1957). *Disc. Faraday Soc.*, **24**, 133–40.

Frank, H. S. (1972). Structural models. In *Water – A Comprehensive Treatise*, vol. 1, ed. F. Franks, pp. 515–43. New York: Plenum Press.

Franks, F. (1975). The hydrophobic interaction. In *Water – A Comprehensive Treatise*, vol. 4, ed. F. Franks, pp. 1–94. New York: Plenum Press.

Franks, F. (1979). Solvent interactions and the solution behaviour of carbohydrates. In *Polysaccharides in Foods*, ed. J. M. V. Blanshard & J. R. Mitchell, pp. 33–50. London: Butterworths.

Franks, F. (1980). Physical, biochemical and physiological effects of low temperatures and freezing – their modification by water soluble polymers. *Scanning Electron Microscopy*, **11**, 349–60.

Franks, F. (1981a). The hydrologic cycle: turnover, distribution and utilization of water. In *Handbook of Water Purification*, ed. W. Lorch, pp. 3–24. Maidenhead: McGraw-Hill.

Franks, F. (1981b). Biophysics and biochemistry of low temperature and freezing. In *Effects of Low Temperatures on Biological Membranes*, ed. G. J. Morris & A. Clarke, pp. 3–20. London: Academic Press.

Franks, F. (1981c). The nucleation of ice in undercooled aqueous solutions. *Cryo-Letters*, **2**, 27–31.

Franks, F. (1982a). Physiological water stress. In *Biophysics of Water*, ed. Franks, F. & S. F. Mathias, pp. 279–94. Chichester: John Wiley & Sons.

Franks, F. (1982b). Water activity as a measure of biological viability and quality control. *Cereal Foods World*, 403–7.

Franks, F. (1982c). The properties of aqueous solutions at subzero temperatures. In *Water – A Comprehensive Treatise*, vol. 7, ed. F. Franks, pp. 215–338. New York: Plenum Press.

Franks, F. (1983a). Water solubility and sensitivity – hydration effects. In *Chemistry and Technology of Water-Soluble Polymers*, ed. C. A. Finch, pp. 157–78. New York: Plenum Press.

Franks, F. (1983b). Solute–water interactions: Do polyhydroxy compounds alter the properties of water? *Cryobiology*, **20**, 335–45.

Franks, F. (1985). Complex aqueous systems at subzero temperatures. In *Properties of Water in Foods in Relation to Quality and Stability*, ed. D. Simatos & J. L. Multon. The Hague: Martinus Nijhoff.

Franks, F., Asquith, M. H., Hammond, C. C., Skaer, H. leB. & Echlin, P. (1977). Polymeric cryoprotectants in the preservation of biological ultrastructure. I. Low temperature states of aqueous solutions of hydrophilic polymers. *J. Microsc.* **110**, 223–38.

Franks, F., Asquith, M. H., Skaer, H. leB. & Roberts, B. (1979). Aggregation patterns and microstructure in aqueous polymer solutions: comparison of quench-fracture electron microscopy with predictions based on rheological measurements. *Cryo-Letters*, **1**, 104–13.

Franks, F. & Eagland, D. (1975). Role of solvent interactions in protein conformation. *Crit. Rev. Biochem.*, **3**, 165–219.

Franks, F., Mathias, S. F., Galfre, P., Webster, S. D. & Brown, D. (1983). Ice nucleation and freezing in undercooled cells. *Cryobiology*, **20**, 298–309.

Franks, F., Mathias, S. F. & Trafford, K. (1984). The nucleation of ice in undercooled water and aqueous polymer solutions. *Colloids Surfaces* **11**, 275–85.

Franks, F. & Morris, E. R. (1978). Blood glycoprotein from antarctic fish. Possible conformational origin of antifreeze activity. *Biochim. Biophys. Acta*, **540**, 346–56.

Franks, F. & Pedley, M. D. (1981). Microcalorimetric study of ternary mixtures of urea and hydrophobic species. *J. Chem. Soc. Faraday Trans.*, 1, **77**, 1341–9.

Franks, F. & Reid, D. S. (1973). Thermodynamic properties. In *Water – A Comprehensive Treatise*, vol. 2, ed. F. Franks, pp. 323–80. New York: Plenum Press.

Friedman, H. L. & Krishnan, C. V. (1973). Thermodynamics of ion hydration. In *Water – A Comprehensive Treatise*, vol. 3, ed. F. Franks, pp. 1–118. New York: Plenum Press.

Frost, A. A. & Pearson, R. G. (1953). *Kinetics and Mechanism*. New York: John Wiley & Sons.

Fuchigami, L. H., Weiser, C. J., Kobayashi, K., Timmis, R. & Gusta, L. V. (1982). A degree growth stage (°GS) model and cold acclimation in temperate woody plants. In *Plant Cold Hardiness and Freezing Stress*, vol. 2, ed. P. H. Li & A. Sakai, pp. 93–116. New York: Academic Press.

Fujikawa, S. (1981). The effect of different cooling rates on the membrane of frozen human erythrocytes. In *Effects of Low Temperatures on Biological Membranes*, ed. G. J. Morris & A. Clarke, pp. 323–34. London: Academic Press.

Geiger, A., Rahman, A. & Stillinger, F. H. (1979). Molecular dynamics study of the hydration of Lennard–Jones solutes. *J. Chem. Phys.*, **70**, 263–76.

Gekko, K. & Morikawa, T. (1981). Preferential hydration of bovine serum albumin in polyhydric alcohol–water mixtures. *J. Biochem., Tokyo*, **90**, 39–50.

George, M. F., Becwar, M. R. & Burke, M. J. (1982). Freezing avoidance by deep undercooling of tissue water in winter-hardy plants. *Cryobiology*, **19**, 628–39.

George, M. F., Burke, M. J., Pellett, H. M. & Johnson, A. G. (1974). Low temperature exotherms and woody plant distribution. *Hort. Science*, **9**, 519–22.

Goldammer, E. v. & Zeidler, M. D. (1969). Molecular motion in aqueous mixtures with organic liquids by NMR relaxation measurements. *Ber. Bunsenges. Phys. Chem.*, **73**, 4–15.

Gould, G. W. & Measures, J. C. (1977). Water relations in single cells. *Phil. Trans. R. Soc. Ser. B*, **278**, 151–66.

Graham, D. & Patterson, B. D. (1982). Responses of plants to low, non-freezing temperatures: proteins, metabolism and acclimation. *A. Rev. Pl. Physiol.*, **33**, 347–72.

Griffiths, J. B. & Beldon, I. (1978). Assessment of cellular damage during cooling to −196 °C using radiochemical markers. *Cryobiology*, **15**, 391–402.

Gusta, L. V., Burke, M. J. & Kappor, A. C. (1975). Determination of unfrozen water in winter cereals at subfreezing temperatures. *Plant Physiology*, **56**, 707–9.

Guthohrlein, G. & Knappe, J. (1968). Structure and function of carbamoyl phosphate synthase. I. Transitions between two catalytically inactive forms and the active form. *Europ. J. Biochem.*, **7**, 119–27.

Guthrie, C., Nashimoto, H. & Nomuro, M. (1969). Structure and function of *E. coli* ribosomes. *VIII.* Cold-sensitive mutants defective in ribosome assembly. *Prod. Natn. Acad. Sci. USA*, **63**, 384–92.

Hart, J. W. & Sabnis, D. D. (1977). Microtubules. In *The Molecular Biology of Plant Cells*, ed. H. Smith, pp. 160–81. Oxford University Press.

Hasted, J. B. & Shahidi, M. (1976). The low frequency dielectric constant of supercooled water. *Nature*, **262**, 777–8.

Hatano, S. & Kabata, K. (1981). Transition of lipid metabolism in relation to frost hardiness in *Chlorella ellipsoidea*. In *Plant Cold Hardiness and Freezing Stress*, ed. P. H. Li & A. Sakai, pp. 145–67. New York: Academic Press.

Hauser, H. (1975). Lipids. In *Water – A Comprehensive Treatise*, vol. 4, ed. F. Franks, pp. 209–304. New York: Plenum Press.

Heber, U., Schmitt, J. M., Krause, G. H., Klossen, R. J. & Santarius, K. A. (1981). Freezing damage to thylakoid membranes *in vitro* and *in vivo*. In *Effects of Low Temperatures on Biological Membranes*, ed. G. J. Morris & A. J. Clarke, pp. 263–84. London: Academic Press.

Hepler, L. G. & Woolley, E. M. (1973). Hydration effects and acid–base equilibria. In *Water – A Comprehensive Treatise*, vol. 3, ed. F. Franks, pp. 145–72. New York: Plenum Press.

Herbert, R. A. (1981). Low temperature adaptation in bacteria. In *Effects of Low Temperatures on Biological Membranes*, ed. G. J. Morris & A. Clarke, pp. 41–54. London: Academic Press.

Hertz, H. G. (1973). Nuclear magnetic relaxation spectroscopy. In *Water – A Comprehensive Treatise*, vol. 3, ed. F. Franks, pp. 301–400. New York: Plenum Press.

von Hippel, P. H. & Hamabata, A. (1973). Model studies on the effects of neutral salts on the conformational stability of biological macromolecules. *J. Mechanochemistry Cell Motility*, **2**, 127–38.

Hirsch, K. R. & Holzapfel, W. B. (1984). Symmetrical hydrogen bonds in ice-X. *Phys. Lett.*, **A101**, 142–4.

Hobbs, P. V. (1974). *Ice Physics*. Oxford University Press.

Hofstee, B. F. (1949). The activation of urease. *J. Gen Physiol.*, **32**, 339–47.

International Institute of Refrigeration (1972). *Recommendations for the Processing and Handling of Frozen Foods*. Paris.

Jackson, H. B. & Cronan, J. E. (1978). An estimate of the minimum amount of fluid lipid required for the growth of *E. coli. Biochim. Biophys. Acta*, **512**, 472–9.

Jacobsen, A. & Pegg, D. E. (1984). Cryopreservation of organs: a review. *Cryobiology*, **21**, 377–84.

Jaenicke, R. (ed.) (1980). *Protein Folding*. Amsterdam: Elsevier North-Holland.

Jaenicke, R. (1981). Enzymes under extremes of physical conditions. *Q. Rev. Biophysics*, **10**, 1–67.

Jarabak, J., Seeds, A. E. & Talalay, P. (1966). Reversible cold inactivation of a 17β-hydroxy-steroid dehydrogenase of human placenta; protective effects of glycerol. *Biochemistry*, **5**, 1269–78.

Jarell, H. C., Butler, K. W., Byrd, R. A., Deslauriers, R., Ekiel, L. & Smith, I. C. P. (1982). A ²H NMR study of *Acholeplasma laidlawii* membranes highly enriched in myristic acid. *Biochim. Biophys. Acta*, **688**, 622–36.

Jeffrey, G. A. (1973). Conformational studies in the solid state: Extrapolation to molecules in solution. *Adv. Chem., Ser.* **117**, 177–96.

Joiner, C. H. & Lauf, P. K. (1979). Temperature dependence of active K^+ transport in cation dimorphic sheep erythrocytes. *Biochim. Biophys. Acta*, **552**, 540–5.

Jones, G. & Dole, M. (1929). The viscosity of aqueous solutions of strong electrolytes with special reference to barium chloride. *J. Amer. Chem. Soc.*, **51**, 2950–64.

Juntilla, O. & Stushnoff, C. (1977). Freezing avoidance by deep supercooling in hydrated lettuce seeds. *Nature*, **269**, 325–7.

Kamb, B. & Prakash, A. (1974). Unpublished result, quoted in P. V. Hobbs, *Ice Physics*, p. 78. Oxford University Press.

Kaplan, J. G., Duphill, M. & Lacroute, F. (1967). A study of the aspartate transcarbamylase activity of yeast. *Arch. Biochem. Biophys.*, **119**, 541–51.

Katz, J. L. & Spaepen, F. (1978). A kinetic approach to nucleation in condensed systems. *Phil. Mag.*, **B37**, 137–48.

Kauzmann, W. (1959). Some factors in the interpretation of protein denaturation. *Adv. Protein Chem.*, **14**, 1–64.

Kell, G. S. (1972). Thermodynamic and transport properties of water. In *Water – A Comprehensive Treatise*, vol. 1, ed. F. Franks, pp. 363–412. New York: Plenum Press.

Kern, C. W. & Karplus, M. (1972). The water molecule. In *Water – A Comprehensive Treatise*, vol. 1, ed. F. Franks, pp. 21–91. New York: Plenum Press.

King, M. W. & Roberts, E. H. (1979). *The Storage of Recalcitrant Seeds – Achievements and Possible Approaches.* Rome: International Board for Plant Genetic Resources.

Kiovsky, T. E. & Pincock, R. E. (1966). Mutarotation of glucose in frozen aqueous solutions. *J. Am. chem. Soc.*, **88**, 7704–10.

Kirkman, H. N. & Hendrickson, E. M. (1962). Glucose 6-phosphate dehydrogenase from human erythrocytes. *J. Biol. Chem.*, **237**, 2371–6.

Konicek, J. & Wadsö, I. (1971). Thermochemical properties of some carboxylic acids, amines and N-substituted amides in aqueous solution. *Acta Chem. Scand.*, **25**, 1541–51.

Kozak, J. J., Knight, W. S. & Kauzmann, W. (1968). Solute–solute interactions in aqueous solutions. *J. Chem. Phys.*, **48**, 675–90.

Kreshek, G. C. (1975). Surfactants. In *Water – A Comprehensive Treatise*, vol. 4, ed. F. Franks, pp. 95–168. New York: Plenum Press.

Krog, J. O., Zachariassen, K. E., Larsen, B. & Smidsrød, O. (1979). Thermal buffering in Afro-alpine plants due to nucleating agent–induced water freezing. *Nature*, **282**, 300–1.

Krug, R. R., Hunter, W. G. & Grieger, R. A. (1976). Enthalpy–entropy compensation. 1. Some fundamental statistical problems associated with the analysis of van't Hoff and Arrhenius data. *J. Phys. Chem.*, **80**, 2335–41.

Lauffer, M. A. (1978). Entropy-driven polymerization of proteins: Tobacco mosaic virus protein and other proteins of biological importance. In *Physical Aspects of Protein Interactions*, ed. N. Catsimpoolas, pp. 115–70. New York. Elsevier North-Holland.

Leibo, S. P. (1980). Water permeability and its activation energy of fertilized and unfertilized mouse ova. *J. Membrane Biol.*, **53**, 179–88.

Leistner, L., Rödel, W. & Krispien, K. (1981). Microbiology of meat and meat products in high- and intermediate-moisture foods. In *Water Activity: Influences on Food Quality*, ed. L. B. Rockland & G. F. Stewart, pp. 855–916. New York: Academic Press.

Leiter, H., Patil, K. J. & Hertz, H. G. (1983). Search for hydrophobic association between small aprotic solutes from an application of the nuclear magnetic relaxation method. *J. Solution Chem.*, **12**, 503–18.

Levitt, J. (1980). *Responses of Plants to Environmental Stresses*, vol. 1. New York: Academic Press.

Levitt, M. (1980). Computer studies of protein molecules. In *Protein Folding*, ed. R. Jaenicke, pp. 17–36. Amsterdam: Elsevier North-Holland.

Lewis, G. N. & Randall, M. (1961). *Thermodynamics*, 2nd edn. New York: McGraw-Hill.

Lian, Y.-N., Chen, A.-T., Suurkuusk, J. & Wadsö, I. (1982). Polyol–water interactions as reflected by aqueous heat capacity values. *Acta Chem. Scand.*, **A36**, 735–9.

Lilley, T. H. (1973). Raman spectroscopy of aqueous electrolyte solutions. In *Water – A Comprehensive Treatise*, vol. 3, ed. F. Franks, pp. 265–300. New York: Plenum Press.

Lindow, S. E. (1983). The role of bacterial ice nucleation in frost injury to plants. *Ann. Rev. Phytopathol.*, **21**, 363–84.

Ling, G. N. (1979). The polarized multilayer theory and other facets of the association–induction hypothesis concerning the distribution of ions and other solutes in living cells. In *The Aqueous Cytoplasm*, ed. A. D. Keith, pp. 23–60. New York: Marcel Dekker.

Livingstone, G., Franks, F. & Aspinall, L. J. (1977). The effects of aqueous solvent structure on the mutarotation of glucose. *J. Solution Chem.*, **6**, 203–16.

Lovelock, J. E. & Bishop, M. W. H. (1959). Prevention of freezing damage to living cells by dimethyl sulphoxide. *Nature*, **183**, 1394–5.

Lyons, J. M. & Raison, J. K. (1970a). Oxidative ability of mitochondria isolated from plant tissues sensitive and resistant to chilling injury. *Plant Physiol.*, **45**, 386–9.

Lyons, J. M. & Raison, J. K. (1970b). A temperature-induced transition in mitochondrial oxidation: Contrasts between cold and warm-blooded animals. *Comp. Biochem. Physiol.*, **37**, 405–11.

Lyons, J. M., Raison, J. K. & Steponkus, P. L. (1979). The plant membrane in response to low temperature. In *Low Temperature Stress in Crop Plants: The Role of the Membrane*, ed. J. M. Lyons, D. Graham & J. K. Raison, pp. 1–24. New York: Academic Press.

MacFarlane, D. R., Kadiyala, R. K. & Angell, C. A. (1983). Homogeneous nucleation and growth of ice from solutions. TTT curves, the nucleation rate and the stable glass criterion. *J. Chem. Phys.*, **79**, 3921–7.

McIntyre, J. A., Gilula, N. B. & Karnovsky, M. J. (1974). Cryoprotectant induced redistribution of intramembranous particles in mouse lymphocytes. *J. Cell Biol.*, **60**, 192–203.

MacKenzie, A. P. (1975). The physico-chemical environment during the freezing and thawing of biological materials. In *Water Relations of Foods*, ed. R. B. Duckworth, pp. 477–504. London: Academic Press.

MacKenzie, A. P. (1977). Nonequilibrium freezing behaviour of aqueous systems. *Phil. Trans. R. Soc. Ser. B*, **278**, 167–88.

MacKenzie, A. P. & Rasmussen, D. H. (1972). Interactions in the water–polyvinyl pyrrolidone system at low temperatures. In *Water Structure at the Water–Polymer Interface*, ed. H. H. G. Jellinek, pp. 146–72. New York: Plenum Press.

McLellan, M. R., Morris, G. J., Coulson, G. E., James, E. R. & Kalinina, L. V. (1984). Role of cytoplasmic proteins in cold shock injury of *Amoeba. Cryobiology*, **21**, 44–59.

McMillan, W. G. & Mayer, J. E. (1945). The statistical thermodynamics of multicomponent mixtures. *J. Chem. Phys.*, **13**, 276–305.

McMurdo, A. C. & Wilson, J. M. (1980). Chilling injury and Arrhenius plots. *Cryo-Letters*, **1**, 231–8.

Martin, R. G. (1963). The first enzyme in histidine biosynthesis: the nature of feedback inhibition by histidine. *J. Biol. Chem.*, **238**, 257–68.

Mathias, S. F., Franks, F. & Trafford, K. (1984). Nucleation and growth of ice in deeply undercooled erythrocytes. *Cryobiology*, **21**, 123–32.

Mayer, E. & Brüggeller, P. (1983). Devitrification of glass water. Evidence for a discontinuity of state? *J. Phys. Chem.*, **87**, 4744–9.

Mazur, P. (1953). Kinetics of water loss from cells at subzero temperature and the likelihood of intracellular freezing. *J. Gen. Physiol.*, **47**, 347–69.

Mazur, P. (1966). Physical and chemical basis of injury in single-celled microorganisms subjected to freezing and thawing. In *Cryobiology*, ed. H. T. Meryman, pp. 214–316. New York: Academic Press.

Mazur, P. (1970). Cryobiology: the freezing of biological systems. *Science*, **168**, 939–49.

Mazur, P., Leibo, S. P. & Miller, R. H. (1974). Permeability of the bovine red cell to glycerol in hyperosmotic solutions at various temperatures. *J. Membrane Biol.*, **15**, 107–36.

Mazur, P., Miller, R. H. & Leibo, S. P. (1974). Survival of frozen-thawed bovine red cells as a function of the permeation of glycerol and sucrose. *J. Membrane Biol.*, **15**, 137–58.

Meryman, H. T. (1966). Review of biological freezing. In *Cryobiology*, ed. H. T. Meryman, pp. 1–114. New York: Academic Press.

Meryman, H. T. (1974). Freezing injury and its prevention in living cells. *Ann. Rev. Biophys. Bioeng.*, **3**, 341–59.

Michelmore, R. W. & Franks, F. (1982). Nucleation rates of ice in undercooled water and aqueous solutions of polyethylene glycol. *Cryobiology*, **19**, 163–71.

Moor, H. & Mühlethaler, K. (1963). Fine structure in frozen-etched yeast cells. *J. Cell. Biol.* **17**, 609–28.

Morris, G. J. (1981). Liposomes as a model system for investigating freezing injury. In *Effects of Low Temperatures on Biological Membranes*, ed. G. J. Morris & A. Clarke, pp. 241–62. London: Academic Press.

Morris, G. J. & Clarke, A. (eds.) (1981). *Effects of Low Temperatures on Biological Membranes*. London: Academic Press.

Morris, G. J., Coulson, G., Meyer, M. A., McLellan, M. R., Fuller, B. J., Grout, B. W. W., Pritchard, H. W. & Knight, S. C. (1983). Cold shock – a widespread cellular reaction. *Cryo-Letters*, **4**, 179–92.

Muhr, A. H. (1983). 'The influence of polysaccharides on ice formation in sucrose solutions.' Unpublished Ph.D. thesis, University of Nottingham.

Nakashima, K., Rudolph, F. B., Wakabayashi, T. & Lardy, H. A. (1975). Rat liver pyruvate carboxylase, V. Reversible dissociation by chloride salts of monovalent cations. *J. Biol. Chem.*, **250**, 331–6.

Narten, A. H. & Levy, H. A. (1972). Liquid water: scattering of X-rays. In *Water – A Comprehensive Treatise*, vol. 1, ed. F. Franks, pp. 311–32. New York: Plenum Press.

Narten, A. H., Venkatesh, C. G. & Rice, S. A. (1976). Diffraction pattern and structure of amorphous solid water at 10 and 77 K. *J. Chem. Phys.*, **64**, 1106–21.

Nichols, N., Sköld, R., Spink, C., Suurkuusk, J. & Wadsö, I. (1976). Additive relations for the heat capacities of non-electrolytes in aqueous solution. *J. Chem. Thermodyn.*, **8**, 1081–93.

Nojima, H., Ikai, A., Oshima, T. & Noda, H. (1977). Reversible unfolding of thermostable phosphoglycerate kinase. *J. Mol. Biol.*, **116**, 429–42.

North, A. C. T. (1979). Functional classification of proteins. In *Characterization of Protein Conformation and Function*, ed. F. Franks, pp. 1–18. London: Symposium Press.

O'Donovan, G. A. & Ingraham, J. L. (1965). Cold sensitive mutants of *E. coli* resulting from increased feedback inhibition. *Proc. Natn. Acad. Sci. USA*, **54**, 451–7.

O'Donovan, G. A. & Neuhard, J. (1970). Pyrimidine metabolism in microorganisms. *Bacteriol. Review*, **34**, 278–343.

Okamoto, B. Y., Wood, R. H. & Thompson, J. (1978). Freezing points of aqueous alcohols. *J. Chem. Soc. Faraday Trans.* 1, **74**, 1990–2007.

Ono, T. A. & Murata, N. (1981). Chilling susceptibility of the blue-green alga *Anacystis nidulans*. II. Stimulation of the passive permeability of cytoplasmic membrane at chilling temperatures. *Plant Physiol.*, **67**, 182–7.

Öquist, G. (1983). Effects of low temperature on photosynthesis. *Plant, Cell and Environment*, **6**, 281–300.

Öquist, G. & Martin, B. (1980). Inhibition of photosynthetic electron transport and formation of inactive chlorophyll in winter stressed *Pinus silvestris*. *Physiol. Plant*, **48**, 33–8.

Packer, K. J. (1977). The dynamics of water in heterogeneous systems. *Phil. Trans. R. Soc. Ser. B*, **278**, 59–86.

Pain, R. H. (1979). The conformational stability of folded proteins. In *Characterization of Protein Conformation and Function*, ed. F. Franks, pp. 19–36. London: Symposium Press.

Parodi, R. M., Bianchi, E. & Ciferri, A. (1973). Thermodynamics of unfolding of lysozyme in aqueous alcohol solutions. *J. Biol. Chem.*, **218**, 4047–51.

Pashley, R. & Israelachvili, J. N. (1981). The long range nature of the hydrophobic interaction. *Colloids Surfaces*, **2**, 169–87.

Penefsky, H. S. & Warner, R. C. (1965). Partial resolution of the enzymes catalyzing oxidative phosphorylation. VI. *J. Biol. Chem.*, **240**, 4696–702.

Petsko, G. A. (1975). Protein crystallography at subzero temperatures: Cryo-protective mother liquors for protein crystals. *J. Mol. Biol.*, **96**, 381–92.

Pfeil, W. & Privalov, P. L. (1976). Thermodynamic investigations of proteins. *Biophys. Chem.*, **4**, 23–50.

Pfeil, W. & Privalov, P. L. (1979). Conformational changes in proteins. In *Biochemical Thermodynamics*, ed. M. N. Jones, pp. 75–115. Amsterdam: Elsevier.

Polge, C., Smith, A. U. & Parkes, A. S. (1949). Revival of spermatozoa after vitrification and dehydration at low temperatures. *Nature*, **164**, 166.

Pratt, L. R. & Chandler, D. (1977). The theory of the hydrophobic effect. *J. Chem. Phys.*, **67**, 3683–704.

Precht, H., Christophersen, J., Hensel, H. & Larcher, W. (1973). *Temperature and Life*. Berlin: Springer.

Pringle, M. J. & Chapman, D. (1981). Biomembrane structure and effects of temperature. In *Effects of Low Temperatures on Biological Membranes*, ed. G. J. Morris & A. Clarke, pp. 21–40. London: Academic Press.

Privalov, P. L. & Khechinashvili, N. N. (1974). A thermodynamic approach to the problem of stabilization of globular protein structure: A calorimetric study. *J. Mol. Biol.*, **86**, 665–84.

Rahman, A. & Stillinger, F. H. (1973). Hydrogen bond patterns in liquid water. *J. Amer. Chem. Soc.*, **95**, 7943–8.

Rasmussen, D. H. (1982). Energetics of homogeneous nucleation. Approach to a physical spinodal. *J. Crystal Growth*, **56**, 45–55.

Rasmussen, D. H. & MacKenzie, A. P. (1972). Effects of solutes on ice-solution interfacial free energy; calculation from measured homogeneous nucleation temperatures. In *Water Structure at the Water–Polymer Interface*, ed. H. H. G. Jellinek, pp. 126–45. New York: Plenum Press.

Reid, D. S. (1979). The low temperature phase behaviour of aqueous ribose. *Cryo-Letters*, **1**, 35–8.

Richards, F. M. & Karplus, M. (1980). Discussion comments. *Biophys. J.*, **32**, 45.

Riehle, U. (1968). 'Über die Vitrifizierung verdünnterwässriger Lösungen.' Unpublished Ph.D. Thesis, Eidesgenössische Technische Hochschule, Zürich.

Robertson, R. E. & Sugamori, S. E. (1972). The hydrolysis of *t*-butyl chloride in aquo-organic mixtures. Heat capacity of activation and solvent structure. *Canad. J. Chem.*, **50**, 1353–60.

Ross, H. K. (1954). Cryoscopic studies. Concentrated solutions of hydroxy compounds. *Ind. Eng. Chem.*, **46**, 601–10.

Rossky, P. J. & Zichi, D. A. (1982). Molecular vibrations and solvent orientational correlations in hydrophobic phenomena. *Faraday Symp. Chem. Soc.*, **17**, 69–78.

Sakai, A., (1974). Characteristics of winter hardiness in extremely hardy twigs. *Fiziol. Rast.*, **21**, 141–7.

Santarius, K. A. (1982). The mechanism of cryoprotection of biomembrane systems by carbohydrates. In *Plant Cold Hardiness and Freezing Stress*, vol. 2, ed. P. H. Li & A. Sakai, pp. 475–86. New York: Academic Press.

Sceats, M. G. & Rice, S. A. (1982). Amorphous solid water and its relationship to liquid water: a random network model for water. In *Water – A Comprehensive Treatise*, vol. 7, ed. F. Franks, pp. 83–214. New York: Plenum Press.

Schnell, R. C. & Vali, G. (1972). Atmospheric ice nuclei from decomposing vegetation. *Nature*, **236**, 163–5.

Schwartz, G. J. & Diller, K. R. (1983*a*). Osmotic response of individual cells during freezing. 1. Experimental volume measurements. *Cryobiology*, **20**, 61–77.

Schwartz, G. J. & Diller, K. R. (1983*b*). Osmotic response of individual cells during freezing. II. Membrane permeability analysis. *Cryobiology*, **20**, 542–52.

Shepard, M. L., Goldston, C. S. & Cocks, F. H. (1976). The H_2O–$NaCl$–glycerol phase diagram and its application in cryobiology. *Cryobiology*, **13**, 9–23.

Shrake, A. & Rupley, J. A. (1973). Environment and exposure to solvent of protein atoms in lysozyme and insulin. *J. Mol. Biol.*, **79**, 351–71.

Silvares, O. M., Cravalho, E. G., Toscano, W. M. & Huggins, C. E. (1975). The thermodynamics of water transport from biological cells during freezing. *Trans. Amer. Soc. Mech. Eng.* pp. 582–8.

Simon, E. W. (1979). Seed germination at low temperatures. In *Low Temperature Stress in Crop Plants: The Role of the Membrane*, ed. J. M. Lyons, D. Graham & J. K. Raison, pp. 37–45. New York: Academic Press.

Simon, E. W. (1981). The low temperature limit for growth and germination. In *Effects of Low Temperatures on Biological Membranes*, ed. G. J. Morris & A. Clarke, pp. 173–88. London: Academic Press.

Singer, S. J. & Nicholson, G. L. (1972). The fluid mosaic model of the structure of cell membranes are viewed as two-dimensional solutions of oriental globular proteins and lipids. *Science*, **175**, 720–31.

Skaer, H. leB., Franks, F., Asquith, M. H. & Echlin, P. (1977). Polymeric cryoprotectants in the preservation of biological ultrastructure. III. Morphological aspects. *J. Microsc.*, **110**, 257–70.

Skaer, H. leB., Franks, F. & Echlin, P. (1978). Non-penetrating polymeric cryofixatives for ultrastructural and analytical studies of biological tissues, *Cryobiology*, **15**, 589–602.

Slayter, R. O. & Morrow, P. A. (1977). Altitudinal variation in the photosynthetic characteristics of snow gum, *Eucalyptus pauciflora*. Sieb. ex Sprend. 1. Seasonal changes under field conditions in the Snowy Mountains area of south-eastern Australia. *Aust. J. Bot.*, **25**, 1–20.

Smith, A. U. (1970). *Current Trends in Cryobiology*. New York: Plenum Press.

Soesanto, T. & Williams, M. C. (1981). Volumetric interpretation of viscosity for concentrated and dilute sugar solutions. *J. Phys. Chem.*, **85**, 3338–41.

Sømme, L. & Conradi-Larsen, E. M. (1977). Cold hardiness of collembolans and orbatid mites from windswept mountain ridges. *Oikos*, **29**, 118–26.

Steponkus, P. L., Dowgert, M. F. & Gordon-Kamm, W. J. (1983). Destabilization of the plasma membrane of isolated plant protoplasts during a freeze–thaw cycle: The influence of cold acclimation. *Cryobiology*, **20**, 448–65.

Storey, K. B. (1983). Metabolism and bound water in overwintering insects. *Cryobiology*, **20**, 365–79.

Storey, K. B., Baust, J. G. & Storey, J. M. (1981). Intermediate metabolism during low temperature acclimation in the overwintering gall fly larva, *Eurosta solidaginis*. *J. Comp. Physiol.*, **144**, 183–90.

Suggett, A. (1975). Polysaccharides. In *Water – A Comprehensive Treatise*, vol. 4, ed. F. Franks, pp. 519–68. New York: Plenum Press.

Synon, M. E. (1979). Life after death in a test tube. *The Daily Telegraph*, September 15, 1979. *See also* Editorial in *Cryo-Letters*, **1**, 77–8 (1979).

Tanford, C. (1970). Protein denaturation. Part C. Theoretical models for the mechanism of denaturation. *Adv. Protein Chem.*, **24**, 1–97.

Taylor, A. O., Slack, C. R. & McPherson, H. G. (1974). Plants under climatic stress. VII. Chill/light. *Plant Physiol.*, **54**, 696–701.

Taylor, M. J. (1979). Assignment of standard pH values to buffers in 20 and 30% (w/w) dimethyl sulfoxide/water mixtures at normal and subzero temperatures. *J. Chem. Eng. Data*, **24**, 230–3.

Taylor, M. J. (1981). The meaning of pH at low temperatures,. *Cryo-Letters*, **2**, 231–9.

Thompson, L. U. & Fennema, O. (1971). Effects of freezing on oxidation of l-ascorbic acid. *J. Agric. Food Chem.*, **19**, 121–9.

Thompson, P. A. (1970). Characterization of the germination response to temperature of species and ecotypes. *Nature*, **225**, 827–31.

Timasheff, S. N. (1978). Thermodynamic examination of the self-association of brain tubulin to microtubules and other stuctures. In *Physical Aspects of Protein Interactions*, ed. N. Catsimpoolas, pp. 219–73. New York: Elsevier North-Holland.

Turnbull, D. (1969). Under what conditions can a glass be formed? *Contemp. Phys.*, **10**, 473–88.

Turnbull, D. & Fisher, J. C. (1949). Rate of nucleation in condensed systems. *J. Chem. Phys.*, **17**, 71–3.

Vonnegut, B. (1947). The nucleation of ice formation by silver iodide. *J. Appl. Phys.*, **18**, 593–5.

Vonnegut, B. & Chessin, H. (1971). Ice nucleation by coprecipitated silver bromide. *Science*, **174**, 945–6.

Whittingham, D. G. (1980). Principles of embryo preservation. In *Low Temperature Preservation in Medicine and Biology*, ed. M. J. Ashwood-Smith & J. Farrant, pp. 65–84. Tunbridge Wells: Pitman Medical Ltd.

Williams, R. J., Willemot, C. & Hope, H. J. (1981). The relationship between cell injury and osmotic volume reduction. IV. The behavior of hardy wheat membrane lipids in monolayer. *Cryobiology*, **18**, 146–54.

Wolfe, J. (1978). Chilling injury in plants – the role of membrane lipid fluidity. *Plant Cell Envir.*, **1**, 241–7.

Wolfe, J. & Steponkus, P. L. (1983). Tension in the plasma membrane during osmotic contraction. *Cryo-Letters*, **4**, 315–22.

Woodcock, A. H., Thistle, M. W., Cook, W. H. & Gibbon, N. E. (1941). The ability of sheep's erythrocytes to survive freezing. *Canad. J. Res.*, **19D**, 206–12.

Zachariassen, K. E. (1980). The role of polyols and nucleating agents in cold-hardy beetles. *J. Comp. Physiol.*, **140**, 227–34.

Index